The Little Prover

The Little Prover

Daniel P. Friedman
Carl Eastlund

Drawings by Duane Bibby

Foreword by J Strother Moore
Afterword by Matthias Felleisen

The MIT Press
Cambridge, Massachusetts
London, England

© 2015 Massachusetts Institute of Technology

All rights reserved. No part of this book may be reproduced in any form by any electronic or mechanical means (including photocopying, recording, or information storage and retrieval) without permission in writing from the publisher.

MIT Press books may be purchased at special quantity discounts for business or sales promotional use. For information, please e-mail special_sales@mitpress.mit.edu.

This book was set in Computer Modern by the authors using LaTeX. Printed and bound in the United States of America.

Library of Congress Cataloging-in-Publication Data

Friedman, Daniel P.
The little prover / Daniel P. Friedman and Carl Eastlund; drawings by Duane Bibby.
p. cm.
Includes bibliographical references and index.
ISBN 978-0-262-52795-8 (pbk. : alk. paper)
1. Automatic theorem proving. 2. LISP (Computer program language) I. Eastlund, Carl II. Title.

QA76.9.A96F745 2015
511.3′6028563—dc23

2015001271

10 9 8 7 6 5 4 3 2

To Mary, Robert, Shannon, Rachel, Sara, Samantha, Brooklyn, Chase, Aria, and to the memory of Brian (1969–1993).

To Mom, Dad, Sharon, Paul Ragnar, Grace, Claire, Paul Duncan, and Ruth.

((**Contents**)

(Foreword **ix**)

(Preface **xi**)

(((1. Old Games, New Rules) **2**) (Examples **181**))

(((2. Even Older Games) **14**) (Examples **182**))

(((3. What's in a Name?) **32**) (Proofs **183**))

(((4. Part of This Total Breakfast) **42**) (Proofs **184**))

(((5. Think It Over, and Over, and Over) **58**) (Proofs **185**))

(((6. Think It Through) **76**) (Proofs **187**))

(((7. Oh My, Stars!) **88**) (Proofs **188**))

(((8. Learning the Rules) **106**) (Proofs **192**))

(((9. Changing the Rules) **114**) (Proofs **193**))

(((10. The Stars Are Aligned) **138**) (Proofs **196**))

((A. Recess) **164**)

((B. The Proof of the Pudding) **180**)

((C. The Little Assistant) **202**)

((D. Restless for More?) **216**)

(Afterword **221**)

(Index **222**))

Foreword

In 1971 Bob Boyer and I began working on an automatic theorem prover for the programming language LISP. In my PhD dissertation of 1973, I wrote:

> This paper describes an automatic theorem prover which is capable of producing inductive proofs of a large number of interesting theorems about functions written in a subset of pure LISP. The program was designed to prove theorems in the way a good programmer might intuit them. It has several features which make it distinct from other systems concerned with proof of program properties: It is fully automatic, requiring no information from the user except the LISP definitions of the functions involved and the theorem to be proved. It automatically uses structural induction when necessary, and automatically generates its own induction formulas. It will occasionally generalize the theorem to be proved, and in so doing, often 'discovers' interesting lemmas. Finally, it is capable of writing new, recursive LISP functions to help properly generalize a theorem.

How is mathematical logic used to describe the behavior of functions or programs? What is a "theorem?" How do you "prove" theorems? What is an "automatic theorem prover?" What do all these mathematical terms—"induction," "generalization," "lemma"—have to do with programs being "correct?" How can "programmer's intuition" guide a foray into mathematics? How do you think about programs that call themselves without just going in circles?

These are some of the questions Bob and I had to answer when we began writing the program described above. Our answers are still evolving 40 years later and are currently manifested in ACL2, the "living" descendant of the prover above.

But while our code is well documented and available for study, the answers to the above questions are not always explicit in it. Sometimes, the best way to learn how to do something is just to sit down and try to do it.

There are two problems with that advice and the problems are especially acute when mathematical logic is involved. First, you have to understand the "rules of the game." Those rules—if followed exactly—will ensure that what you "prove" is really true. Second, it is hard to keep in mind the precise statement of every rule—and if you make a mistake you might end up believing something is true when it is not.

This little book and the accompanying little assistant addresses both of these problems. The book itself provides you with a gentle introduction to the mathematics behind all of this. It presents "the rules" in a way geared toward the programmer. Indeed the rules themselves are largely intuitive to the programmer. Second, the assistant enforces the rules for you. But unlike the "automatic theorem provers" that I create, their assistant is designed to enforce the rules while allowing you to learn by doing.

The offer here is: the prover will make sure the "formulas" are formulas and the "proofs" are proofs. But you're the intelligent actor: so what do you do now?

J Strother Moore
Austin, Texas

Preface

What does it mean for a statement to be *true*? Some statements can be verified directly. To determine whether a particular omelette is delicious, we merely have to taste the omelette. Our answer is imprecise, however. We must wonder: how tasty must a "delicious" omelette be? What egg dishes are properly called "omelettes?" Assuming our taste test succeeds, we can answer the question for only one omelette at a time. We may never know whether all omelettes are delicious, even if each individual omelette we try is tasty.

What does it mean for a statement to be true of a recursive function? It is easy to test an individual case. When we evaluate `(reverse (reverse '(1 2)))`, we reach the result `'(1 2)`, exactly as we expect. Furthermore, evaluating a recursive function follows a predictable—and fortunately, often simple—set of rules. We can therefore answer more general questions. For instance, does `(reverse (reverse x))` always produce `x` for any list? We can determine the answer without ever evaluating the expression or even knowing the specific value of `x`.

Our goal is to teach the reader how to determine facts about recursive functions using induction. Our approach is to start with programming concepts such as recursive functions and lists, and to lead the reader along the shortest path to inductive proofs. We aim to teach enough to verify simple properties such as whether `(reverse (reverse x))` always produces `x` for any list, although we leave that particular example as an exercise for the reader.

Understanding how to read, write, and evaluate recursive functions over lists is adequate preparation. We assume knowledge of neither logic nor mathematics beyond arithmetic. We express as much as we can using simple programming concepts.

Acknowledgements

We have many people to thank for their contributions to this book. Thanks, as always, to Dorai Sitaram for the SLaTeX typesetting program. Thanks to Jared Davis, whose dissertation on the theorem prover Milawa has been a useful reference for the core of ACL2's logic.

For help with ACL2 and the implementation of `dethm`, we thank Pete Manolios and Matt Kaufmann. We are also grateful to J Moore and Bob Boyer for their contributions to theorem proving research and for allowing us to name our little proof assistant, J-Bob, after them.

We thank our mentors and inspirations for teaching and guiding us: Will Byrd, Matthias Felleisen, Bob Filman, George Springer, and Mitch Wand. We especially want to thank Matthias for forging our partnership.

Thanks to our early supporters: Adam Foltzer, Ron Garcia, Amr Sabry, Christian Urban, and Yin Wang, and everyone else who has been kind and helpful. Thanks to Jason Hemann, who has given us endless help, support, and feedback on drafts of the book and the implementation of J-Bob. For their feedback on drafts, thanks to Conrad Barski, Daniel Brady, Will Byrd, Kyle Carter, Josh Cox, Jim Duey, Matthias Felleisen,

Bob Filman, Adam Foltzer, Ron Garcia, Jaime Guerrero, Jason Hemann, Joe Hendrix, Andrew Kent, J Moore, Joe Near, Rex Page, Paul Snively, Vincent St-Amour, Dylan Thurston, Christian Urban, Dale Vaillaincourt, Michael Vanier, and Dave Yrueta.

Thanks to our illustrator, Duane Bibby, for illustrations, for inspiration, and for coming through with our last-minute requests. Special thanks to our editor, Marie Lufkin Lee, for helping us throughout the process of bringing this book to its final form.

We also thank our hosts for the many visits to Bloomington, where most of our work took place. Thanks to our friends Katie Edmonds and Sam Tobin-Hochstadt for hosting one such visit, and our eternal gratitude to Mary Friedman for hosting all of the others, for graciously tolerating our constant postponements and demands on her time, and for supporting this book every step of the way.

Notation

Expressions in the book are written in a language comprised of variables, quoted literals, if expressions, and function applications. Functions include nine built-in operators as well as user-defined functions, which may be recursive.

Variable names consist of one or more characters, including letters, numbers, and most punctuation; notable exceptions are parentheses and apostrophe. Examples of variables are x, +, and variable-name1.

Quoted literals are preceded by a quote symbol, written ', and include atomic "symbols" such as 'banana, natural numbers such as '12, and lists such as '(), '(a glass of orange juice), and '(bacon with 2 eggs). Lists may also be arbitrarily deeply nested; for example, '((3 slices of toast) or (1 bagel with cream cheese)).

An if expression has three parts: the *question*, the *answer*, and the *else*. For example, (if sleepy 'coffee '(orange juice)) produces 'coffee if sleepy is 't and produces '(orange juice) if sleepy is 'nil.

Function applications include a function name and zero or more arguments. For example, (cons x '(with hash browns)), (f x (g y z)), or (do-something).

The nine built-in functions are: cons, which adds an element to the front of a list; car, which returns the first element of a non-empty list; cdr, which returns the tail of a non-empty list excluding its first element; atom, which returns 'nil for non-empty lists and 't for everything else; equal, which returns 't if its arguments have identical values and 'nil otherwise; natp, which returns 't if its argument is a natural number and 'nil otherwise; size, which counts the conses needed to build a value; +, which adds two natural numbers; and <, which returns 't if its first argument is less than its second argument and 'nil otherwise.

User-defined functions are written with defun, and include the function's name, a list of names for its arguments, and an expression for the function's body. Function definitions may be recursive. For example:

```
(defun list-length (xs)
  (if (atom xs)
      '0
      (+ '1 (list-length (cdr xs)))))
```

Theorems, introduced in chapter 1, are defined with dethm. Much like functions, theorem definitions include the theorem's name, a list of names for its arguments, and an expression for its body. Since expressions cannot refer to theorems, theorems are not recursive. Here is an example:

```
(dethm natp/list-length (xs)
  (natp (list-length xs)))
```

Guidelines for the Reader

For those who wish to "play along," we include a simple proof assistant, J-Bob, defined in the same language as the theorems we prove. J-Bob is a program that can check each step when attempting to prove a theorem but does *not* contribute any steps. In the appendices, we introduce J-Bob in "Recess," we present the complete code of every example and proof in the book using J-Bob in "The Proof of the Pudding," and we include the implementation of J-Bob in "The Little Assistant." J-Bob, support to run J-Bob in several languages, and the aforementioned proofs are available for download at `http://the-little-prover.org/`.

Food appears in some examples for two reasons. First, food is easier to visualize than abstract symbols (but there will be many short symbols like x, etc.). We hope the food imagery helps you to better understand the examples and concepts. Second, we want to provide a little distraction. We know how exhausting the subject matter can be, thus these breakfast foods are for energizing you. As such, we hope that thinking about food will cause you to occasionally set the book aside and have a bite.

You are now ready to start. Good luck! We hope you enjoy the book.

Bon appétit!

Daniel P. Friedman
Bloomington, Indiana

Carl Eastlund
Brooklyn, New York

The Little Prover

1.
Old Games, New Rules

1 Salutations.

What are salutations?

2 Salutations are a fancy way of saying hello or good morning.†

†Thank you, E. B. White (1899-1985), for *Charlotte's Web.*

Good morning.

3 Have you read *The Little Schemer*?

'nil.

4 Oh, so you've read *The Little LISPer*?

Well, . . .

5 Do you remember "Cons the Magnificent?"

Certainly.†

†Otherwise, continue if you have some familiarity with recursion.

6 What is `(car (cons 'ham '(eggs)))` *equal* to?

'ham.†

†We write all values as expressions using the symbol ' to denote a literal "quoted" value, rather than referring to values outside of expressions.

7 Yes. But also `(car (cons 'ham '(cheese)))`, `(car (cdr (cons 'eggs '(ham))))`, `(car (cons (car '(ham)) '(eggs)))`, . . .

That's strange.

8 What *value* is this expression equal to?

```
(car (cons 'ham '(eggs)))
```

That's easy.

```
'ham
```
†

†When we rewrite one expression to another that is equal to it, we put them side-by-side. They are smaller to accommodate much larger expressions.

Exactly.

9 Are there any others?

No,
an expression is only equal to one value.

10 Tricky.

What value is this expression equal to?

```
(atom '())
```

11 That's easy, too.

```
't
```

What value is this equal to?

```
(cons a b)
```

12 We do not know what a and b are.

Does that mean (cons a b) has no value?

13 Until we know what a and b are, we do not know what value (cons a b) is equal to.

Can we find a value for this expression?

```
(atom (cons 'ham '(eggs)))
```

14 Of course.

```
'nil
```

How about this expression?

```
(atom (cons a b))
```

15 We still do not know what a and b are.

Nevertheless, we can figure out what value it has. Try again.

```
(atom (cons a b))
```

16 'nil,
because no matter what values the variables a and b have, cons cannot produce an atom.

```
'nil
```

Does this expression have a value?

(equal 'flapjack (atom (cons a b)))

17 Well, we already know that (atom (cons a b)) is equal to 'nil.

How can we use that?

18 If we can replace (atom (cons a b)) with 'nil on its own, then surely we can replace the (atom (cons a b)) in (equal 'flapjack (atom (cons a b))) with 'nil.

In other words, we want to *focus* on (atom (cons a b)) in the *context* of the outer equal expression.[†]

(equal 'flapjack (atom (cons a b)))

[†] We show this by writing the focus in black and its context in blue.

19 So, does that mean we can replace this focus with 'nil?

(equal 'flapjack 'nil)

Precisely. In that case, what value is (equal 'flapjack 'nil) equal to?

(equal 'flapjack 'nil)

20 'nil,
of course.

'nil

What value is (equal 'flapjack (atom (cons a b))) equal to?

21 'nil,
as we have just seen.

How many steps did we take to get from (equal 'flapjack (atom (cons a b))) to 'nil?

22 Two.

What is the first step?

23 In the first step, the focus (atom (cons a b)) is equal to 'nil.

What is the whole expression?	24 (equal 'flapjack (atom (cons a b))).
Where is this focus in the whole expression?	25 It is the second argument to equal.
What is the second step?	26 In the second step, the whole expression is equal to 'nil.
Where is the focus in the whole expression?	27 The focus *is* the whole expression.
What value is this expression equal to? (atom (cdr (cons (car (cons p q)) '())))	28 We don't know what p and q are, but perhaps we can find a value anyway.
What is the first step? (atom (cdr (cons (car (cons p q)) '())))	29 The car of (cons p q) is always equal to p, regardless of p and q. (atom (cdr (cons p '())))
And what is the second step? (atom (cdr (cons p '())))	30 Of course, the cdr of (cons p '()) is always equal to '(), regardless of p. (atom '())
And finally? (atom '())	31 We know that (atom '()) is 't. 't
That took three steps. Can we do it in fewer?	32 We are up to the challenge.

How shall we start?

(atom (cdr (cons (car (cons p q)) '())))

33 The cdr of (cons (car (cons p q)) '()) is always equal to '(), regardless of p and q.

(atom '())

We have seen this step before.

(atom '())

34 And so we are done.

't

How many *axioms* have we used?

35 What is an axiom?

An axiom is a basic assumption that is presumed to be true. For one example, we assume that (atom (cons x y)) is always equal to 'nil. For another, we assume that (car (cons x y)) is always equal to x. We also assume that (cdr (cons (car (cons x y)) '())) is always equal to '(). Finally, we assume that (cdr (cons x '())) is always equal to '().

36 Then we have used four axioms.

Can we rephrase the third and fourth assumptions more generally?

37 Yes,
the cdr of (cons x y) is always equal to y. Does this mean we have used only three axioms?

Yes. Shall we view our axioms so far?

38 We are excited to see them.

The Axioms of Cons (initial)

```
(dethm atom/cons (x y)
  (equal (atom (cons x y)) 'nil))
```

```
(dethm car/cons (x y)
  (equal (car (cons x y)) x))
```

```
(dethm cdr/cons (x y)
  (equal (cdr (cons x y)) y))
```

Now that we have names for the axioms, we can use them again and again.

39 What does dethm mean?

It means *de*fine a *th*eore*m*.

40 What is a theorem?

A theorem is an expression that is always true. When we use dethm, we also include a list of the variables used in the expression.

41 What is the difference between an axiom and a theorem?

Axioms are theorems that are assumed to be true, whereas other theorems must be shown to be true.

42 What does equal mean?

The function equal tells us whether two values are equal. What is the value of this expression?

```
(equal 'eggs '(ham))
```

43 Its value is 'nil, because 'eggs is not equal to '(ham).

```
'nil
```

---

44 Exactly. What is the value of this expression?

```
(car
  (cons (equal (cons x y) (cons x y))
    '(and crumpets)))
```

It has the same value as (car (cons 't '(and crumpets))), because (cons x y) is always equal to (cons x y), regardless of x and y.

```
(car
  (cons 't
    '(and crumpets)))
```

---

45 And, of course, the second step is easy.

```
(car (cons 't '(and crumpets)))
```

Delicious!

```
(car '(t and crumpets))
```

---

46 Is this a theorem?

```
(car '(t and crumpets))
```

But, of course!

```
't
```

---

47 What is the value of this expression?

```
(equal (cons x y) (cons 'bagels '(and lox)))
```

We do not know. It depends on the values of x and y.

---

48 What else is this expression equal to?

```
(equal (cons x y) (cons 'bagels '(and lox)))
```

Perhaps it is equal to many things.

---

49 Does the order of the arguments to equal matter?

```
(equal (cons x y) (cons 'bagels '(and lox)))
```

No, (cons x y) is equal to (cons 'bagels '(and lox)) in the same cases that (cons 'bagels '(and lox)) is equal to (cons x y).

```
(equal (cons 'bagels '(and lox)) (cons x y))
```

---

50 Exactly.

It sounds like we have some new axioms.

The Axioms of Equal (initial)

```
(dethm equal-same (x)
  (equal (equal x x) 't))
```

```
(dethm equal-swap (x y)
  (equal (equal x y) (equal y x)))
```

What is different about equal-swap compared to the other axioms we have seen thus far?

51 In the other axioms, the second argument of the outer equal is shorter than the first argument. In equal-swap, neither argument of the outer equal is shorter than the other.

Does it matter?

52 Yes, we think so.

That's kind of true. It is useful to know how an axiom can simplify an expression. For the same reason that equal-swap is true, however, we could write the axioms in either order without changing their meaning.

53 How fascinating.

What is this focus equal to, according to car/cons?

```
(cons y
  (equal (car (cons (cdr x) (car y)))
         (equal (atom x) 'nil)))
```

54 The car of (cons (cdr x) (car y)) is (cdr x).

```
(cons y
  (equal (cdr x)
         (equal (atom x) 'nil)))
```

What else is this focus equal to, according to car/cons? Recall that "is equal to" works in both directions.

```
(cons y
  (equal (car (cons (cdr x) (car y)))
         (equal (atom x) 'nil)))
```

55 In that case, (car (cons (cdr x) (car y))) is equal to many things according to car/cons, such as this focus.

```
(cons y
  (equal (car (cons
               (car (cons (cdr x) (car y)))
               '(oats)))
         (equal (atom x) 'nil)))
```

Can we use atom/cons here?

```
(cons y
  (equal (car (cons (car (cons (cdr x) (car y)))
                    '(oats)))
         (equal (atom x)
                'nil)))
```

56 Indeed, we can use atom/cons to replace 'nil with many different expressions.

```
(cons y
  (equal (car (cons (car (cons (cdr x) (car y)))
                    '(oats)))
         (equal (atom x)
                (atom
                  (cons (atom (cdr (cons a b)))
                    (equal (cons a b) c))))))
```

What is this focus equal to?

```
(cons y
  (equal (car (cons (car (cons (cdr x) (car y)))
                    '(oats)))
         (equal (atom x)
                (atom
                  (cons (atom (cdr (cons a b)))
                    (equal (cons a b) c))))))
```

57 According to cdr/cons, it is equal to simply b.

```
(cons y
  (equal (car (cons (car (cons (cdr x) (car y)))
                    '(oats)))
         (equal (atom x)
                (atom
                  (cons (atom b)
                    (equal (cons a b) c))))))
```

Are there any axioms we have not yet used on the example beginning in frame 54?

58 Yes, equal-same and equal-swap.

Can we use either of them here?

```
(cons y
  (equal (car (cons (car (cons (cdr x) (car y)))
                    '(oats)))
         (equal (atom x)
                (atom
                  (cons (atom b)
                    (equal (cons a b) c))))))
```

59 Yes, equal-swap.

```
(cons y
  (equal (car (cons (car (cons (cdr x) (car y)))
                    '(oats)))
         (equal (atom x)
                (atom
                  (cons (atom b)
                    (equal c (cons a b)))))))
```

What value is this expression equal to?

```
(cons y
  (equal (car (cons (car (cons (cdr x) (car y)))
                    '(oats)))
         (equal (atom x)
                (atom
                  (cons (atom b)
                    (equal c (cons a b)))))))
```

60 That is a good question. We do not know, but we have had fun playing with it so far!

The Law of Dethm (initial)

For any theorem (dethm *name* ($x_1 \ldots x_n$) $body_x$), the variables $x_1 \ldots x_n$ in $body_x$ can be replaced with any corresponding expressions $e_1 \ldots e_n$. The result, $body_e$, can be used to rewrite a focus p to become q provided $body_e$ is either (equal p q) or (equal q p).

---

61 Let's try one more example. In car/cons, what are *name*, x_1, x_2, and $body_x$ from the Law of Dethm?

```
(dethm car/cons (x y)
  (equal (car (cons x y)) x))
```

The axiom's *name* is car/cons, x_1 and x_2 are x and y, respectively, and $body_x$ is (equal (car (cons x y)) x).

---

62 To rewrite this focus using car/cons, what expressions should we use for e_1 and e_2 from the Law of Dethm?

(atom (car (cons (car a) (cdr b))))

We use (car a) as e_1 and (cdr b) as e_2.

---

63 In that case, how do we figure out $body_e$ from the Law of Dethm based on $body_x$?

If we replace x with (car a) and replace y with (cdr b), then $body_e$ is (equal (car (cons (car a) (cdr b))) (car a)).

---

64 Have we found p and q?

(atom (car (cons (car a) (cdr b))))

Given $body_e$, p is (car (cons (car a) (cdr b))) and q is (car a). Since this focus *is* p, we may replace it with q.

(atom (car a))

---

65 Now work through frames 55-59 again using the Law of Dethm for each one.

Sounds challenging.

If this gets too challenging, we have an assistant named J-Bob that can help.	66	Who is J-Bob?
J-Bob is a program that helps us rewrite one expression to another. J-Bob "knows" about all the axioms and the Law of Dethm and makes sure we get all the details right.	67	J-Bob certainly sounds helpful.
We can meet J-Bob on page 164, and play along with J-Bob for all of the examples in this chapter on page 181.	68	Must we meet J-Bob to continue reading?
Absolutely not, but the deeper we go, the more J-Bob can help.	69	And so we shall make a visit.
That's probably a good idea. Before we head over there, perhaps we should fortify ourselves with two helpings of our favorite breakfast.	70	Certainly.

2 - Even Older Games

1 What is this focus *obviously* equal to?

```
(if (car (cons a b))
  c
  c)
```

a, since the car of (cons a b) is always a.

```
(if a
  c
  c)
```

2 What axiom tells us this obvious fact?

car/cons, that's easy.

3 What is this expression *obviously* equal to?

```
(if a
  c
  c)
```

The result of this if is c, regardless of a. So perhaps the expression is equal to c. But we do not know any axioms that tell us this.

4 Perhaps we need some axioms about if.

If only.

The Axioms of If (initial)

```
(dethm if-true (x y)
  (equal (if 't x y) x))
```

```
(dethm if-false (x y)
  (equal (if 'nil x y) y))
```

```
(dethm if-same (x y)
  (equal (if x y y) y))
```

5 What is this expression *obviously* equal to?

```
(if a
  c
  c)
```

c, by if-same.

```
c
```

What else is c equal to, according to if-same?

6 Doesn't if-same require an if expression?

If if-same can start with an if expression and end with a variable, then it *must* also be able to start with a variable and end with an if expression. So... what else is c equal to, according to if-same?

```
c
```

7 How about this?

```
(if (if (equal a 't)
        (if (equal 'nil 'nil)
            a
            b)
        (equal 'or (cons 'black '(coffee))))
    c
    c)
```

Absolutely!

8 Are there any other such if expressions?

We can fill in any if question we want, as long as we keep c as the if answer and the if else.

9 What are the if question, if answer, and if else?

Every if expression has three parts: (if Q A E). We call them the if question, the if answer, and the if else, or Q, A, and E for short.

10 Very well.

What value is this focus equal to?

```
(if (if (equal a 't)
        (if (equal 'nil 'nil)
            a
            b)
        (equal 'or (cons 'black '(coffee))))
    c
    c)
```

11 '(black coffee), which certainly helps us focus.

```
(if (if (equal a 't)
        (if (equal 'nil 'nil)
            a
            b)
        (equal 'or '(black coffee)))
    c
    c)
```

12 Can we simplify the innermost if question?

```
(if (if (equal a 't)
        (if (equal 'nil 'nil)
            a
            b)
        (equal 'or '(black coffee)))
    c
    c)
```

Certainly, using equal-same.

```
(if (if (equal a 't)
        (if 't
            a
            b)
        (equal 'or '(black coffee)))
    c
    c)
```

13 Do we have to use equal-same to replace (equal 'nil 'nil) with 't?

No, we can always use equal itself, because 'nil is a value.

14 Can we simplify the innermost if now?

```
(if (if (equal a 't)
        (if 't
            a
            b)
        (equal 'or '(black coffee)))
    c
    c)
```

Yes, by if-true.

```
(if (if (equal a 't)
        a
        (equal 'or '(black coffee)))
    c
    c)
```

15 Does the if question (equal a 't) tell us anything about this focus?

```
(if (if (equal a 't)
        a
        (equal 'or '(black coffee)))
    c
    c)
```

Yes, since this focus is in the if answer of the question (equal a 't), we know that the focus a is equal to 't.

16 Can we use that knowledge to rewrite the focus in frame 15?

Presumably. But what axiom allows us to replace a with 't?

17 We need a new axiom about if and equal.

We are eager to see it.

The Axioms of Equal (final)

```
(dethm equal-same (x)
  (equal (equal x x) 't))
```

```
(dethm equal-swap (x y)
  (equal (equal x y) (equal y x)))
```

```
(dethm equal-if (x y)
  (if (equal x y) (equal x y) 't))
```

Which axiom allows us to rewrite a to 't in frame 15?

18 We assume the answer is the new axiom, equal-if. But how does this axiom work? We cannot use the Law of Dethm, because the body of the axiom is not an application of equal.

Then we must revise The Law of Dethm.

19 Interesting.

The Law of Dethm (final)

For any theorem (dethm *name* (x_1 ... x_n) $body_x$), the variables x_1 ... x_n in $body_x$ can be replaced with any corresponding expressions e_1 ... e_n. The result, $body_e$, can be used to rewrite a focus as follows:

1. $body_e$ must contain the *conclusion* (equal p q) or (equal q p),
2. the conclusion must not be found in the question of any if or in the argument of any function application,
3. and if the conclusion can be found in an if answer (respectively else), then the focus must be found in an if answer (respectively else) with the same question.

In `equal-if`, what are *name*, x_1, x_2, and $body_x$ from the Law of Dethm?

20 The axiom's *name* is `equal-if`, x_1 and x_2 are `x` and `y`, and $body_x$ is `(if (equal x y) (equal x y) 't)`.

In order to rewrite the focus of frame 15 using `equal-if`, what expressions should we use for e_1 and e_2 from the Law of Dethm?

21 We use `a` as e_1 and `'t` as e_2.

In that case, how do we figure out $body_e$ from the Law of Dethm based on $body_x$?

22 If we use `a` for `x` and `'t` for `y`, $body_e$ is `(if (equal a 't) (equal a 't) 't)`.

What expression contained in $body_e$ do we use as our *conclusion*?

23 Our conclusion is `(equal a 't)`, since we are rewriting `a` to `'t`.

The expression `(equal a 't)` is contained in $body_e$ twice, in the `if` question and in the `if` answer. Which one is our conclusion?

24 The expression in the `if` answer is our conclusion, since according to the Law of Dethm the conclusion must not be found in the question of any `if`.

Is the conclusion found in an `if` answer?

25 Yes, it is found in the answer of an `if` with the question `(equal x y)`.

Is the focus in frame 15 also found in an `if` answer with the `if` question `(equal x y)`?

26 Yes, it is.

Is the conclusion found in an `if` else?

27 No, it is not.

Then according to the Law of Dethm, we can rewrite the focus a to become 't.

```
(if (if (equal a 't)
        a
        (equal 'or '(black coffee)))
    c
    c)
```

28 Finally. But why is that expression *orange*?

```
(if (if (equal a 't)
        't
        (equal 'or '(black coffee)))
    c
    c)
```

We write expressions in orange to draw attention. Here, (equal a 't) is a premise that allows us to rewrite the focus†.

†Orange expressions may be part of a context or a focus, even though they are not blue or black. In frame 28, the orange expression is in a context. In frame 43 of the next chapter, the orange expression is in a focus.

29 What is a premise?

A *premise* is an if question such that a focus can be found in either the if answer or the if else. In frame 28, why is (equal a 't) a premise?

30 In frame 28, (equal a 't) is a premise because the focus is in the if answer.

Let's walk through another example using premises and the final Law of Dethm.

31 That's a good idea.

Here is a new dethm.

```
(dethm jabberwocky† (x)
  (if (brillig x)
      (if (slithy x)
          (equal (mimsy x) 'borogove)
          (equal (mome x) 'rath))
      (if (uffish x)
          (equal (frumious x) 'bandersnatch)
          (equal (frabjous x) 'beamish))))
```

†Thank you, Lewis Carroll (1832-1898).

32 Is that really a theorem?

Perhaps, depending on what brillig, slithy, mimsy, mome, uffish, frumious, and frabjous mean. For now it is just an example, and we will *pretend* it is a theorem. Can we use jabberwocky to rewrite this focus?

```
(cons 'gyre
  (if (uffish '(callooh callay))
     (cons 'gimble
       (if (brillig '(callooh callay))
          (cons 'borogove '(outgrabe))
          (cons 'bandersnatch '(wabe))))
     (cons (frabjous '(callooh callay)) '(vorpal))))
```

33

Perhaps. This is a complicated expression, and jabberwocky is a complicated dethm.

In order to use jabberwocky, we must find an equal expression that matches this focus. Do any of the equal expressions in jabberwocky have arguments that are similar to (frabjous '(callooh callay))?

34

Yes, the first argument of the last equal expression in jabberwocky is (frabjous x).

What must we substitute for x to make (frabjous x) equal to the focus?

35

We must substitute '(callooh callay) for x.

Exactly. If we substitute '(callooh callay) for x in jabberwocky, we get $body_e$, which we created to meet the first condition of the Law of Dethm.

```
(if (brillig '(callooh callay))
   (if (slithy '(callooh callay))
      (equal (mimsy '(callooh callay)) 'borogove)
      (equal (mome '(callooh callay)) 'rath))
   (if (uffish '(callooh callay))
      (equal (frumious '(callooh callay)) 'bandersnatch)
      (equal (frabjous '(callooh callay)) 'beamish)))
```

What *conclusion* must we use to rewrite the focus (frabjous '(callooh callay))?

36

We must use the last equal expression, (equal (frabjous '(callooh callay)) 'beamish), as the conclusion, thus meeting the first condition. And if we meet the second and third conditions, we could rewrite the focus to 'beamish.

Looking at frame 36, is the conclusion found in an if question or in the argument of a function application?

37 No, it is not, thus meeting our second condition.

Still looking at frame 36, is the conclusion found in an if answer?

38 No, it is not.

And still looking at frame 36, is the conclusion found in an if else?

39 Yes, it is found in the else of two ifs, thus meeting the first part of the third condition. So, now do we have to meet the second part of this condition?

Yes. Still looing at frame 36, what are the questions of the ifs that have the conclusion in their else?

40 The questions of the ifs are (brillig '(callooh callay)) and (uffish '(callooh callay)).

Looking at frame 33, is the focus found in the else of any ifs with the questions (brillig '(callooh callay)) and (uffish '(callooh callay))?

41 The focus is in the else of an if with the question (uffish '(callooh callay)), but it is not in the else of an if with the question (brillig '(callooh callay)). So, *no*, it is not.

Then, we can not use jabberwocky to rewrite the focus in frame 33. The conclusion from frame 36 does not meet the second part of the third condition.

42 All that effort, and we still don't get to use jabberwocky.

Let's try again. Can we use jabberwocky to rewrite this focus instead?

```
(cons 'gyre
  (if (uffish '(callooh callay))
      (cons 'gimble
        (if (brillig '(callooh callay))
            (cons 'borogove '(outgrabe))
            (cons 'bandersnatch '(wabe))))
      (cons (frabjous '(callooh callay)) '(vorpal))))
```

43 Perhaps.

Do any of the equal expressions in jabberwocky have arguments that are similar to 'bandersnatch?

44 Yes, the second argument of the third equal expression in jabberwocky is 'bandersnatch.

What must we substitute for x to make 'bandersnatch equal to the focus in frame 43?

45 The focus in frame 43 is already 'bandersnatch.

In that case, we can choose an expression to substitute for x that helps our conclusion satisfy the conditions of the Law of Dethm.

46 How do we do that?

What must we substitute for x to make the *premises* of 'bandersnatch in jabberwocky equal to the *premises* of the focus in frame 43?

47 Once again, we must substitute '(callooh callay) for x.

And once again, substituting '(callooh callay) for x in jabberwocky produces the expression in frame 36. What conclusion must we use this time?

48 The third equal expression, whose arguments are (frumious '(callooh callay)) and 'bandersnatch,

because one of equal's arguments, 'bandersnatch, is equal to the focus in frame 43.

This conclusion must meet the conditions in the Law of Dethm. Is the conclusion found in an if question or in the argument of a function application?

49 No, it is not.

Is the conclusion found in an if answer?

50 Indeed it is, since the conclusion, which contains equal's two arguments: (frumious '(callooh callay)) and 'bandersnatch is found in the answer of the if question (uffish '(callooh callay)).

Is the focus in frame 43 found in the answer of an if with the question (uffish '(callooh callay))?

51 Yes, it is.

Is the conclusion found in an if else?

52 Yes, it is found in the else of an if with the question (brillig '(callooh callay)).

Is the focus in frame 43 found in an if else with the if question (brillig '(callooh callay))?

53 Yes, it is.

In that case, the conclusion in frame 48 meets the conditions in the Law of Dethm. We *can* use jabberwocky to rewrite the focus in frame 43 with the premises written in orange. We can rewrite 'bandersnatch to become (frumious '(callooh callay)).

```
(cons 'gyre
  (if (uffish '(callooh callay))
      (cons 'gimble
        (if (brillig '(callooh callay))
            (cons 'borogove '(outgrabe))
            (cons 'bandersnatch '(wabe))))
      (cons (frabjous '(callooh callay)) '(vorpal))))
```

54 *O frabjous day!*

```
(cons 'gyre
  (if (uffish '(callooh callay))
      (cons 'gimble
        (if (brillig '(callooh callay))
            (cons 'borogove '(outgrabe))
            (cons (frumious '(callooh callay)) '(wabe))))
      (cons (frabjous '(callooh callay)) '(vorpal))))
```

How many more times should we read the jabberwocky example?

55 That's easy. Exactly as many times as needed until it is completely understood. But why do we use a pretend theorem?

Sometimes our intuitions get in the way. Axioms like car/cons and if-same are easy to understand, even without the Law of Dethm. Since jabberwocky means nothing, we must understand the Law of Dethm thoroughly in order to use it.

56 Are there other, real theorems that use premises?

57

Let's see. Does the if question (equal (cdr (car a)) '(hash browns)) tell us anything about this focus?

```
(if (atom (car a))
   (if (equal (car a) (cdr a))
      'hominy
      'grits)
   (if (equal (cdr (car a)) '(hash browns))
      (cons 'ketchup (car a))
      (cons 'mustard (car a))))
```

Not really, since that if question tells us (cdr (car a)) is equal to '(hash browns), and this focus is neither of those expressions.

58

Does the if question (equal (car a) (cdr a)) tell us anything about this focus?

```
(if (atom (car a))
   (if (equal (car a) (cdr a))
      'hominy
      'grits)
   (if (equal (cdr (car a)) '(hash browns))
      (cons 'ketchup (car a))
      (cons 'mustard (car a))))
```

No, since neither the if answer nor the if else of that question contains this focus.

59

Does the if question (atom (car a)) tell us anything about this focus?

```
(if (atom (car a))
   (if (equal (car a) (cdr a))
      'hominy
      'grits)
   (if (equal (cdr (car a)) '(hash browns))
      (cons 'ketchup (car a))
      (cons 'mustard (car a))))
```

Yes, since that focus is in the if else of the question (atom (car a)), we know that (car a) must be a cons with its own car and cdr. Perhaps there is an axiom to this effect.

The Axioms of Cons (final)

```
(dethm atom/cons (x y)
  (equal (atom (cons x y)) 'nil))
```

```
(dethm car/cons (x y)
  (equal (car (cons x y)) x))
```

```
(dethm cdr/cons (x y)
  (equal (cdr (cons x y)) y))
```

```
(dethm cons/car+cdr (x)
  (if (atom x) 't (equal (cons (car x) (cdr x)) x)))
```

60

Can we therefore rewrite this focus using the premise (atom (car a))?

```
(if (atom (car a))
  (if (equal (car a) (cdr a))
    'hominy
    'grits)
  (if (equal (cdr (car a)) '(hash browns))
    (cons 'ketchup
      (car a))
    (cons 'mustard (car a))))
```

We should be able to rewrite (car a) as a cons of its own car and cdr.

```
(if (atom (car a))
  (if (equal (car a) (cdr a))
    'hominy
    'grits)
  (if (equal (cdr (car a)) '(hash browns))
    (cons 'ketchup
      (cons (car (car a)) (cdr (car a))))
    (cons 'mustard (car a))))
```

61

What axiom allows us to rewrite the focus in frame 60?

The axiom cons/car+cdr.

62

Does the if question (equal (cdr (car a)) '(hash browns)) tell us anything about this focus?

```
(if (atom (car a))
  (if (equal (car a) (cdr a))
    'hominy
    'grits)
  (if (equal (cdr (car a)) '(hash browns))
    (cons 'ketchup
      (cons (car (car a)) (cdr (car a))))
    (cons 'mustard (car a))))
```

Yes, since this focus is in the if answer of the question (equal (cdr (car a)) '(hash browns)), we know that (cdr (car a)) is equal to '(hash browns).

Can we use that knowledge as a premise to rewrite this focus?

```
(if (atom (car a))
  (if (equal (car a) (cdr a))
    'hominy
    'grits)
  (if (equal (cdr (car a)) '(hash browns))
    (cons 'ketchup
      (cons (car (car a)) (cdr (car a))))
    (cons 'mustard (car a))))
```

63 We should be able to rewrite (cdr (car a)) as '(hash browns).

```
(if (atom (car a))
  (if (equal (car a) (cdr a))
    'hominy
    'grits)
  (if (equal (cdr (car a)) '(hash browns))
    (cons 'ketchup
      (cons (car (car a)) '(hash browns)))
    (cons 'mustard (car a))))
```

Do we know of any axiom that justifies the rewrite in frame 63?

64 Yes, equal-if.

Do we think there are more axioms that use premises?

65 There must be.

The Axioms of If (final)

```
(dethm if-true (x y)
  (equal (if 't x y) x))
```

```
(dethm if-false (x y)
  (equal (if 'nil x y) y))
```

```
(dethm if-same (x y)
  (equal (if x y y) y))
```

```
(dethm if-nest-A (x y z)
  (if x (equal (if x y z) y) 't))
```

```
(dethm if-nest-E (x y z)
  (if x 't (equal (if x y z) z)))
```

Can the axiom if-same rewrite this expression?

```
(cons 'statement
  (cons
    (if (equal a 'question)
       (cons n '(answer))
       (cons n '(else)))
    (if (equal a 'question)
       (cons n '(other answer))
       (cons n '(other else)))))
```

66 That does not seem likely.

Can we rewrite the previous expression to this one?

```
(cons 'statement
  (if (equal a 'question)
     (cons
       (if (equal a 'question)
          (cons n '(answer))
          (cons n '(else)))
       (if (equal a 'question)
          (cons n '(other answer))
          (cons n '(other else))))
     (cons
       (if (equal a 'question)
          (cons n '(answer))
          (cons n '(else)))
       (if (equal a 'question)
          (cons n '(other answer))
          (cons n '(other else))))))
```

67 Yes we can, using if-same. Apparently if-same can rewrite the expression in the previous frame, after all.

Exactly. How can we use if-same to rewrite the expression from frame 66 to the expression from frame 67?

```
(cons 'statement
  (cons
    (if (equal a 'question)
       (cons n '(answer))
       (cons n '(else)))
    (if (equal a 'question)
       (cons n '(other answer))
       (cons n '(other else)))))
```

68 We use if-same where y is

```
(cons
  (if (equal a 'question)
     (cons n '(answer))
     (cons n '(else)))
  (if (equal a 'question)
     (cons n '(other answer))
     (cons n '(other else)))).
```

and x is (equal a 'question).

```
(cons 'statement
  (if (equal a 'question)
     (cons
       (if (equal a 'question)
          (cons n '(answer))
          (cons n '(else)))
       (if (equal a 'question)
          (cons n '(other answer))
          (cons n '(other else))))
     (cons
       (if (equal a 'question)
          (cons n '(answer))
          (cons n '(else)))
       (if (equal a 'question)
          (cons n '(other answer))
          (cons n '(other else))))))
```

If (equal a 'question) is true in the premise, must it also be true in this focus?

```
(cons 'statement
  (if (equal a 'question)
     (cons
       (if (equal a 'question)
          (cons n '(answer))
          (cons n '(else)))
       (if (equal a 'question)
          (cons n '(other answer))
          (cons n '(other else))))
     (cons
       (if (equal a 'question)
          (cons n '(answer))
          (cons n '(else)))
       (if (equal a 'question)
          (cons n '(other answer))
          (cons n '(other else))))))
```

69 Yes, truly, according to if-nest-A.

```
(cons 'statement
  (if (equal a 'question)
     (cons
       (cons n '(answer))
       (if (equal a 'question)
          (cons n '(other answer))
          (cons n '(other else))))
     (cons
       (if (equal a 'question)
          (cons n '(answer))
          (cons n '(else)))
       (if (equal a 'question)
          (cons n '(other answer))
          (cons n '(other else))))))
```

If (equal a 'question) is false in the premise, must it also be false in this focus?

```
(cons 'statement
  (if (equal a 'question)
     (cons (cons n '(answer))
       (if (equal a 'question)
          (cons n '(other answer))
          (cons n '(other else))))
     (cons
       (if (equal a 'question)
          (cons n '(answer))
          (cons n '(else)))
       (if (equal a 'question)
          (cons n '(other answer))
          (cons n '(other else))))))
```

70 Yes, certainly, according to if-nest-E.

```
(cons 'statement
  (if (equal a 'question)
     (cons (cons n '(answer))
       (if (equal a 'question)
          (cons n '(other answer))
          (cons n '(other else))))
     (cons
       (cons n '(else))
       (if (equal a 'question)
          (cons n '(other answer))
          (cons n '(other else))))))
```

If (equal a 'question) is true (false) in the premise, must it also be true (false) in the focus?

```
(cons 'statement
  (if (equal a 'question)
     (cons (cons n '(answer))
       (if (equal a 'question)
          (cons n '(other answer))
          (cons n '(other else))))
     (cons (cons n '(else))
       (if (equal a 'question)
          (cons n '(other answer))
          (cons n '(other else))))))
```

71 Yes, truly (and certainly), according to if-nest-A (and if-nest-E).

```
(cons 'statement
  (if (equal a 'question)
     (cons (cons n '(answer))
       (cons n '(other answer)))
     (cons (cons n '(else))
       (cons n '(other else)))))
```

What expression do we start with in frame 66?

72

```
(cons 'statement
  (cons
    (if (equal a 'question)
       (cons n '(answer))
       (cons n '(else)))
    (if (equal a 'question)
       (cons n '(other answer))
       (cons n '(other else))))).
```

What expression do we end up with in frame 71?

73

```
(cons 'statement
  (if (equal a 'question)
    (cons (cons n '(answer))
      (cons n '(other answer)))
    (cons (cons n '(else))
      (cons n '(other else))))).
```

How many if expressions are in frame 72?

74 Two.

How many if expressions are in frame 73?

75 One.

Is that interesting?

76 Indeed, it is.

How is J-Bob doing?

77 We still have not met J-Bob.

Now would be an excellent time to take a recess, meet J-Bob on page 164, and play through this chapter on page 182.

78 Maybe this is a good time to give J-Bob a try.

Don't go when tired and hungry.

79 We can have a plate of waffles topped with butter, syrup, and strawberries while we meet J-Bob.

3 - What's in a Name?
Name

1 What value is (pair 'sharp 'cheddar) equal to?

'(sharp cheddar).

2 What value is (first-of (pair 'baby 'swiss)) equal to?

'baby.

3 What value is (second-of (pair 'monterey 'jack)) equal to?

'jack.

4 Are these our favorite omelette ingredients?

Perhaps.

5 Here is the function pair.

```
(defun pair (x y)
  (cons x (cons y '())))
```

Now define first-of and second-of.

No surprises here.

```
(defun first-of (x)
  (car x))
```

And none here, either.

```
(defun second-of (x)
  (car (cdr x)))
```

6

```
(dethm first-of-pair (a b)
  (equal (first-of (pair a b)) a))
```

Is the claim first-of-pair a theorem?

What is a claim?

7 A *claim* is an as-yet unproven theorem.

So far first-of-pair does not appear to be a theorem, since none of the axioms we know at this point can be used here.

Let's try to prove the claim first-of-pair.

8 How do we prove a claim?

We *prove* a claim by writing a proof, of course.

9 And what is a proof?

A *proof* is a sequence of rewriting steps that ends in 't. If we can rewrite a claim, step by step, to 't, then that claim is a theorem.

10 Let's get started.

Since (pair x y) is (cons x (cons y '())), what must (pair a b) be?

11 (cons a (cons b '())).

The Law of Defun (initial)

Given the non-recursive function
(defun *name* $(x_1 \ldots x_n)$ *body*),

(*name* $e_1 \ldots e_n$) = *body* **where** x_1 **is** e_1, …, x_n **is** e_n.

Apply the Law of Defun using the definition of pair.

```
(equal (first-of (pair a b)) a)
```

12 The body of pair is (cons x (cons y '())), and we replace x with a and y with b.

```
(equal (first-of (cons a (cons b '()))) a)
```

What axiom applies here?

```
(equal (first-of (cons a (cons b '()))) a)
```

13 No axiom, but we can apply the Law of Defun using first-of.

```
(equal (car (cons a (cons b '()))) a)
```

Is the proof done yet?

(equal (car (cons a (cons b '()))) a)

14 Yes, once we use car/cons and equal-same.

't

Is first-of-pair a theorem?

15 Yes,
by the Law of Defun, car/cons, and equal-same.

Which function definitions does the proof use?

16 It uses pair and first-of.

```
(dethm second-of-pair (a b)
  (equal (second-of (pair a b)) b))
```

Is second-of-pair a theorem?

17 We shall find out.

Use the definition of second-of.

(equal (second-of (pair a b)) b)

18 We rewrite the focus using the Law of Defun.

(equal (car (cdr (pair a b))) b)

Now use the definition of pair.

(equal (car (cdr (pair a b))) b)

19 Easily done.

(equal (car (cdr (cons a (cons b '())))) b)

What next?

(equal (car (cdr (cons a (cons b '())))) b)

20 Use cdr/cons, then car/cons.

(equal b b)

Correct. Does this focus reduce to 't?

(equal b b)

21 Yes,
by equal-same.

't

Is second-of-pair a theorem?

22 Yes,
by car/cons, cdr/cons, equal-same, second-of, and pair.

In first-of-pair, we used the Law of Defun on pair first, but in second-of-pair, we used the Law of Defun on pair second.

23 Does that order matter?

No,
not in this particular case.

24 Can the order ever matter?

Certainly, depending on the proof.

25 Can we always find a proof, whichever order we choose first?

If we can find a proof one way, we can always find it another. If the second way goes wrong, we can "back up" to where we started and do it the first way again. But some approaches will find a proof faster than others.

26 That is useful to know.

What does in-pair? do?

```
(defun in-pair? (xs)
  (if (equal (first-of xs) '?)
      't
      (equal (second-of xs) '?)))
```

27 The function in-pair? determines whether the two-element list† xs contains '?.

†Of course, in-pair? works on more inputs than just two-element lists. See chapter 4.

We can try to prove this claim.

```
(dethm in-first-of-pair (b)
  (equal (in-pair? (pair '? b)) 't))
```

28 So in other words, prove the claim that in-pair? finds '? when it is the first element of a list?

Yes. Shall we begin?

```
(equal (in-pair? (pair '? b))
       't)
```

29 First we use the definition of pair.

```
(equal (in-pair? (cons '? (cons b '())))
       't)
```

What now?

```
(equal (in-pair? (cons '? (cons b '())))
       't)
```

30 Now we use the definition of in-pair?.

```
(equal (if (equal (first-of (cons '? (cons b '())))
                  '?)
           't
           (equal (second-of (cons '? (cons b '())))
                  '?))
       't)
```

And?

```
(equal (if (equal (first-of (cons '? (cons b '())))
                  '?)
           't
           (equal (second-of (cons '? (cons b '())))
                  '?))
       't)
```

31 Next, we use the definition of first-of.

```
(equal (if (equal (car (cons '? (cons b '())))
                  '?)
           't
           (equal (second-of (cons '? (cons b '())))
                  '?))
       't)
```

Can we simplify further?

```
(equal (if (equal (car (cons '? (cons b '())))
                  '?)
           't
           (equal (second-of (cons '? (cons b '())))
                  '?))
       't)
```

32 Yes, using car/cons and equal-same.

```
(equal (if 't
           't
           (equal (second-of (cons '? (cons b '())))
                  '?))
       't)
```

Our claim looks simple from here.

```
(equal (if 't
           't
           (equal (second-of (cons '? (cons b '())))
                  '?))
       't)
```

33 It certainly does.

```
't
```

Is in-first-of-pair a theorem?

34 Yes,
by car/cons, equal-same, if-true, pair, in-pair?, and first-of.

Here's a similar claim.

```
(dethm in-second-of-pair (a)
  (equal (in-pair? (pair a '?)) 't))
```

35 This time, do we need to prove that in-pair? finds '? when it is the *second* element of a two-element list?

Exactly. Use pair first.

```
(equal (in-pair? (pair a '?))
       't)
```

36 This is a familiar step.

```
(equal (in-pair? (cons a (cons '? '())))
       't)
```

Is this step familiar, too?

```
(equal (in-pair? (cons a (cons '? '())))
       't)
```

37 Yes, here we use in-pair?.

```
(equal (if (equal (first-of (cons a (cons '? '())))
                  '?)
           't
           (equal (second-of (cons a (cons '? '())))
                  '?))
       't)
```

And then?

```
(equal (if (equal (first-of (cons a (cons '? '())))
                  '?)
           't
           (equal (second-of (cons a (cons '? '())))
                  '?))
       't)
```

38 We can use first-of and car/cons in the if question.

```
(equal (if (equal a '?)
           't
           (equal (second-of (cons a (cons '? '())))
                  '?))
       't)
```

Does the if question help in our proof of in-second-of-pair?

39 The if question is (equal a '?), but we don't know whether a is equal to '?.

In that case, let's skip to the if else.

```
(equal (if (equal a '?)
           't
           (equal (second-of (cons a (cons '? '())))
                  '?))
       't)
```

40 Okay. In the if else, we use second-of.

```
(equal (if (equal a '?)
           't
           (equal (car (cdr (cons a (cons '? '()))))
                  '?))
       't)
```

Here are two easy steps.

```
(equal (if (equal a '?)
           't
           (equal (car (cdr (cons a (cons '? '()))))
                  '?))
       't)
```

41 Indeed.

```
(equal (if (equal a '?)
           't
           (equal '? '?))
       't)
```

Obviously, '? is equal to '?.

```
(equal (if (equal a '?)
           't
           (equal '? '?))
       't)
```

42 That is indeed obvious.

```
(equal (if (equal a '?)
           't
           't)
       't)
```

Can we rewrite this if question now?

```
(equal (if (equal a '?)
           't
           't)
       't)
```

43 No, but that is fine. We do not need its value after all.

```
't
```

Insight: Skip Irrelevant Expressions

Rewriting a claim to 't does not have to go in any particular order. Some parts of the expression might be skipped entirely. For example, if-same can simplify many if expressions to 't regardless of the if question.

Now have we proved in-second-of-pair?	44	Yes, by car/cons, cdr/cons, equal-same, if-same, pair, in-pair?, first-of, and second-of.
Did we need to use first-of?	45	No, so our proof should be shorter, right?
Could be. Try out shorter proofs with J-Bob on page 183.	46	Now that we have met J-Bob, we can't wait.
Don't rush to the next chapter. Take some time off and have a healthy snack.	47	Should we read this chapter one more time?
Perhaps.	48	Perhaps we shall, over a bowl of oatmeal, dates, and blueberries.

Part of This
Total Breakfast

What value is (list0? 'oatmeal) equal to?

1 'nil,
because 'oatmeal is not a list.

What value is (list0? '()) equal to?

2 't,
because '() is the empty list.

What value is (list0? '(toast)) equal to?

3 'nil,
because the list '(toast) is not empty.

Define list0?.

4

```
(defun list0? (x)
  (if (equal x 'oatmeal)
      'nil
      (if (equal x '())
          't
          (if (equal x '(toast))
              'nil
              'nil))))
```

Very funny. Try again.

5

```
(defun list0? (x)
  (equal x '()))
```

Is list0? total?

6 What does "total" mean?

"list0? is *total*" means that no matter what value v is passed to list0?, the expression (list0? v) has a value.

7 In that case, list0? is total if equal is total.

The function equal is total.

8 Then list0? is total.

How do we arrive at that answer?

9 The function list0? applies equal to x and '(). As long as equal has a value for all arguments, list0? does as well.

Well put. Take a bow.

10 *Grazie.*

What value is (list1? 'oatmeal) equal to?

11 'nil,
because 'oatmeal is not a list.

What value is (list1? '()) equal to?

12 'nil,
because the list '() does not have exactly one element.

What value is (list1? '(toast)) equal to?

13 't,
because '(toast) is a list of one element.

What value is (list1? '(raisin oatmeal)) equal to?

14 'nil,
because '(raisin oatmeal) is not a one-element list.

Define list1?.

15

```
(defun list1? (x)
  (if (atom x)
      'nil
      (list0? (cdr x))))
```

Is list1? total?

16 We have to think about that. This function is more complicated than list0?.

There's no rush. We can take our time with this.

17 The if question in list1? asks atom of x. Is atom total?

Yes, atom is total.	18	What about if?
An if expression produces either its answer or its else no matter what *value* its question has.	19	In other words, as long as an if's question, answer, and else have values, the if expression does too.
Precisely.	20	Okay. And what about cdr and car? Are they total functions?
Yes, all built-in operators are total.	21	That's surprising!
Strange, but true.	22	What value is (cdr 'grapefruit) equal to?
We are only concerned with the result of cdr on conses; (cdr 'grapefruit) must have a value, but that is all we need to know.	23	What about (cdr '())?
Yes, (cdr '()) has a value.	24	Do we need to know what its value is?
No.	25	All right. What about (car '())?
Same story.	26	We are ready to answer the question in frame 16.
To repeat the question, is list1? total?	27	Yes, because atom, cdr, and list0? are total, and the if's question, answer, and else all have values.

What value is (list2? 'oatmeal) equal to?

28 'nil,
predictably.

What value is (list2? '(hash browns)) equal to?

29 't,
because '(hash browns) is a two-element list.

What value is (list2? '(vinegared hash browns)) equal to?

30 'nil,
because '(vinegared hash browns) is not a two-element list.

Define list2?.

31

```
(defun list2? (x)
  (if (atom x)
      'nil
      (list1? (cdr x))))
```

Is list2? total?

32 Yes,
for all the same reasons that list1? is total.

Do we understand what total functions are now?

33 Perhaps.

In that case, we can update the Law of Defun.

34 Really?

The Law of Defun (final)

Given the total function (defun *name* (x_1 ... x_n) *body*),

(*name* e_1 ... e_n) = *body* **where** x_1 **is** e_1, ..., x_n **is** e_n.

Yes. Now that we know what "total" means, we can use the Law of Defun for any total function, including all non-recursive functions and many recursive functions.

35 But why does it matter if a function is total?

Good question. Our axioms and laws tell us which expressions have equal values. If an expression does not have a value, our axioms and laws do not apply.

36 Can we use our axioms and laws for a function that is not total anyway?

Functions that are not total are called *partial functions*. Here is one.

```
(defun partial (x)
  (if (partial x)
      'nil
      't))
```

What value is (partial 'nil) equal to?

37 It has no value.

We can try to prove this claim.

```
(dethm contradiction ()
  'nil)
```

38 This is a very strange claim.

Nevertheless, let's try to prove that 'nil is equal to 't.

39 We are eager to see how this goes.

Expand the claim using if-same, where x is (partial x) and y is 'nil.

```
'nil
```

40 Simple enough.

```
(if (partial x)
    'nil
    'nil)
```

Expand the if answer below using the premise written in orange and if-nest-A where x is (partial x), y is 'nil, and z is 't.

```
(if (partial x)
    'nil
    'nil)
```

41 That is simple as well.

```
(if (partial x)
    (if (partial x)
        'nil
        't)
    'nil)
```

Now expand the if else similarly, using the orange premise and if-nest-E where x is (partial x), y is 't, and z is 'nil.

```
(if (partial x)
    (if (partial x)
        'nil
        't)
    'nil)
```

42 Very well.

```
(if (partial x)
    (if (partial x)
        'nil
        't)
    (if (partial x)
        't
        'nil))
```

Next, use partial in both focuses.

```
(if (partial x)
    (if (partial x)
        'nil
        't)
    (if (partial x)
        't
        'nil))
```

43 Okay.

```
(if (partial x)
    (if (if (partial x)
            'nil
            't)
        'nil
        't)
    (if (if (partial x)
            'nil
            't)
        't
        'nil))
```

44 Now use if-nest-A and if-nest-E and the premise (partial x) to drop the new ifs.

```
(if (partial x)
  (if (if (partial x)
          'nil
          't)
      'nil
      't)
  (if (if (partial x)
          'nil
          't)
      't
      'nil))
```

Interesting.

```
(if (partial x)
  (if 'nil
      'nil
      't)
  (if 't
      't
      'nil))
```

---

45 Simplify these ifs.

```
(if (partial x)
  (if 'nil
      'nil
      't)
  (if 't
      't
      'nil))
```

Easy. We use if-false and if-true.

```
(if (partial x)
  't
  't)
```

---

46 Now use if-same.

```
(if (partial x)
  't
  't)
```

How inconsistent! We have just shown that 'nil is equal to 't.

```
't
```

---

47 Have we?

We started with 'nil and used if-same, if-false, if-true, if-nest-A, if-nest-E, and partial to rewrite the claim to 't.

---

48 Wrong!

Why?
Isn't this exactly what we did?

---

49 Remember: the Law of Defun only works for *total* functions. And partial is not total.

Does that mean 'nil is not equal to 't?

Exactly right.

50 What a relief!
Otherwise, it would be impossible to imagine this book.
But, aren't there any total functions that we could use to prove contradiction?

There are none.

51 Now it is clear why total functions matter!

Good. Shall we define list3? now?

52 How many more functions like this are there?

We will never run out until we reach the largest list.

53 That will take forever and ever.

Do we know a faster way?

54 Certainly. How about recursion?

Go for it.

55

```
(defun list? (x)
  (if (atom x)
      (equal x '())
      (list? (cdr x))))
```

What does list? do?

56 It produces 't if its argument is a list; that is, either '() or a cons whose cdr is also a list. Otherwise, list? produces 'nil.

Is list? total?

57 We think so. Every value is a list, or it is not a list.†

† We do not consider infinite lists or circular lists; we only consider values that can be constructed by finite programs using the operators we present.

But are we sure?	58	We are not certain.
Look again at the definition of list? in frame 55.	59	In order for the if to have a value, its question, answer, and else must, too.
Go on.	60	We already know that atom is total.
And?	61	The answer of the if is obvious.
But what about the else?	62	For that, we need to know that cdr and list? are total.
Right.	63	Wait a minute!
Tricky, isn't it?	64	To find out whether list? is total, we need to know whether list? is total. But isn't that where we started?
So far, so good. We know that (list? (cdr x)) passes a value to list? that has fewer conses than it started with. And we know that if we do this enough times, we will reach an atom.	65	That makes sense, but it's hardly a proof.
Correct, so let's prove it.	66	What, exactly, is the claim?
Our claim is that the measure of list? decreases on every recursive call.	67	What is a measure?

A *measure* is an expression that is included with a function definition. It may only refer to previously defined, total functions and to the function definition's formal arguments. The measure must produce a natural number that decreases for every recursive call to the function.

68 And what is list?'s measure?

The measure of list? is (size x).

```
(defun list? (x)
  (if (atom x)
      (equal x '())
      (list? (cdr x))))

measure: (size x)
```

69 What is size?

We count the conses in a value using size. What value is (size '((1 (a 2)) b)) equal to?

70 That is easy: '6.

Exactly. What value is (size '((10 (A 20)) B)) equal to?

71 That is also '6.

Precisely. What about (size '(10 (A 20)))?

72 '4.

And (size '10)?

73 That is '0, because there are no conses in '10.

Now we can state the claim that list? is total:

"If x is not an atom, then (size (cdr x)) must be smaller than (size x)."

Is this claim true?

74 Indeed, (cdr x) has been built with one fewer cons than x.

Here is the totality claim:

```
(if (natp (size x))
    (if (atom x)
        't
        (< (size (cdr x)) (size x)))
    'nil)†
```

The function natp tells whether its argument is a *natural number*, meaning 0, 1, 2, ...

†See how this claim is constructed from the definition of list? and its measure in chapter 8.

75 Is there an axiom that tells us that the measure (size x) is a natural number?

The Axioms of Size

```
(dethm natp/size (x)
  (equal (natp (size x)) 't))
```

```
(dethm size/car (x)
  (if (atom x) 't (equal (< (size (car x)) (size x)) 't)))
```

```
(dethm size/cdr (x)
  (if (atom x) 't (equal (< (size (cdr x)) (size x)) 't)))
```

Yes,
and (natp (size x)) states that the measure cannot decrease forever.

Are we ready to prove that list? is total?

76 Ready and willing with these new axioms.

Can we use natp/size?

```
(if (natp (size x))
  (if (atom x)
    't
    (< (size (cdr x)) (size x)))
  'nil)
```

77 And off we go!

```
(if 't
  (if (atom x)
    't
    (< (size (cdr x)) (size x)))
  'nil)
```

One if down.

```
(if 't
  (if (atom x)
    't
    (< (size (cdr x)) (size x)))
  'nil)
```

78 And one to go.

```
(if (atom x)
  't
  (< (size (cdr x)) (size x)))
```

What do we know about comparing sizes using the orange premise below?

```
(if (atom x)
  't
  (< (size (cdr x)) (size x)))
```

79 We know size/cdr.

```
(if (atom x)
  't
  't)
```

And now we are done.

```
(if (atom x)
  't
  't)
```

80 Yes, we are, by if-same.

```
't
```

We have just proved that list? is total.

81 Must we prove that every function is total?

Yes, but since all non-recursive functions have the same measure and totality claim, we can just prove them all total at once.	82 Let's do it.
Here is the claim that a non-recursive function is total. 't	83 Interesting, the claim for a non-recursive function to be total is 't. The proof completes itself!
Now we know that many functions that use recursion on lists can be total.	84 What about other kinds of recursion?
The function sub replaces every '? in its second argument with the value of its first argument. What value is (sub 't '?) equal to?	85 't, because sub replaces '? with 't.
What value is (sub '(a ? b) '(x ? y)) equal to?	86 '(x (a ? b) y), because sub replaces '? in '(x ? y) with '(a ? b).
What value is (sub 'and '(ham (? eggs) ? (toast (?) butter))) equal to?	87 '(ham (and eggs) and (toast (and) butter)).

88 Define sub.

```
(defun sub (x y)
  (if (atom y)
      (if (equal y '?)
          x
          y)
      (cons (sub x (car y))
        (sub x (cdr y)))))

measure: (size y)
```

89 The definition of sub *does* come with a measure. Well done.

We hope that (size y) is the appropriate measure.

90 Here is the claim that sub is total.

```
(if (natp (size y))
   (if (atom y)
      't
      (if (< (size (car y)) (size y))
         (< (size (cdr y)) (size y))
         'nil))
   'nil)
```

We know that (size y) is a natural number.

```
(if (atom y)
   't
   (if (< (size (car y)) (size y))
      (< (size (cdr y)) (size y))
      'nil))
```

91 This proof is going quickly. And the finale?

```
(if (atom y)
   't
   (if (< (size (car y)) (size y))
      (< (size (cdr y)) (size y))
      'nil))
```

size/car and size/cdr, using the premise (atom y).

```
(if (atom y)
   't
   (if 't
      't
      'nil))
```

92 Are there any other kinds of recursion we can try to prove total?

How about the recursion in partial? Are we sure we can't prove that total?

Good idea. Is partial total?

```
(if (natp (size x))
  (if (< (size x) (size x))
    't
    'nil)
  'nil)
```

93 We can take off the outer if, at least.

```
(if (< (size x) (size x))
  't
  'nil)
```

And then what?

94 We do not know. This claim looks false—hopefully (size x) is not less than (size x)!

Exactly. And that is why partial is partial.

95 That's clear.

Time for a recess?

96 J-Bob awaits on page 184.

First let's finish our breakfast.

97 It is the most important meal of the day.

5. Think It Over and Over and Over and Over

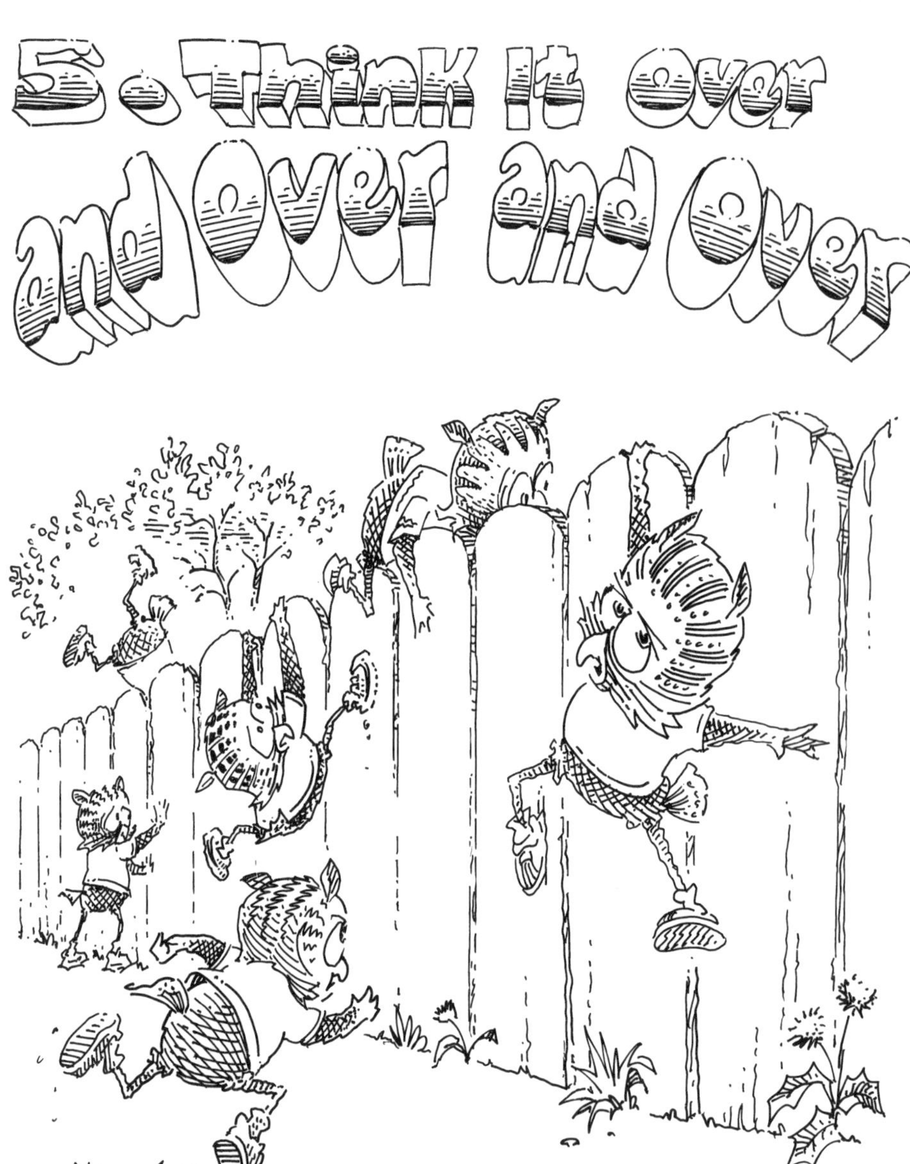

Is memb?[†] familiar?

```
(defun memb? (xs)
  (if (atom xs)
      'nil
      (if (equal (car xs) '?)
          't
          (memb? (cdr xs)))))
```

[†]The name memb? ends with a ? *not* because it searches for '? but because it always returns 't or 'nil.

1 Yes,
it is sort of an old acquaintance, but it is just checking to see if there is a '? in xs.

Is remb familiar, too?

```
(defun remb (xs)
  (if (atom xs)
      '()
      (if (equal (car xs) '?)
          (remb (cdr xs))
          (cons (car xs)
            (remb (cdr xs))))))
```

2 Yes,
this one has also been around for a while. Are these functions total?

Perhaps they are. Here is memb? with a measure.

```
(defun memb? (xs)
  (if (atom xs)
      'nil
      (if (equal (car xs) '?)
          't
          (memb? (cdr xs)))))

measure: (size xs)
```

3 What must we prove?

Here is the claim that memb? is total.

```
(if (natp (size xs))
  (if (atom xs)
    't
    (if (equal (car xs) '?)
      't
      (< (size (cdr xs)) (size xs))))
  'nil)
```

4 We start by using natp/size and if-true.

```
(if (atom xs)
  't
  (if (equal (car xs) '?)
    't
    (< (size (cdr xs)) (size xs))))
```

As in the proof of list?'s totality, size/cdr reduces this focus using the premise (atom xs).

```
(if (atom xs)
  't
  (if (equal (car xs) '?)
    't
    (< (size (cdr xs)) (size xs))))
```

5 And that's that.

```
(if (atom xs)
  't
  (if (equal (car xs) '?)
    't
    't))
```

Here is the measure of remb.

```
(defun remb (xs)
  (if (atom xs)
    '()
    (if (equal (car xs) '?)
      (remb (cdr xs))
      (cons (car xs)
        (remb (cdr xs))))))

measure: (size xs)
```

6 The function size is a versatile measure. What is the totality claim for remb?

Here is the totality claim.

```
(if (natp (size xs))
  (if (atom xs)
    't
    (< (size (cdr xs)) (size xs)))
  'nil)
```

7 Once again, we can start with natp/size and if-true.

```
(if (atom xs)
  't
  (< (size (cdr xs)) (size xs)))
```

And then?

```
(if (atom xs)
  't
  (< (size (cdr xs)) (size xs)))
```

8 size/cdr finishes the job using the premise (atom xs).

```
(if (atom xs)
  't
  't)
```

Can we prove memb?/remb0, which states that remb removes '? from a 0-element list?

```
(dethm memb?/remb0 ()
  (equal (memb? (remb '())) 'nil))
```

9 We can try.

Go right ahead.

```
(equal (memb?
         (remb '()))
       'nil)
```

10 We start by using remb, replacing xs by '().

```
(equal (memb?
         (if (atom '())
           '()
           (if (equal (car '()) '?)
             (remb (cdr '()))
             (cons (car '())
               (remb (cdr '()))))))
       'nil)
```

Can the outer if expression be simplified?

```
(equal (memb?
         (if (atom '())
           '()
           (if (equal (car '()) '?)
             (remb (cdr '()))
             (cons (car '())
               (remb (cdr '()))))))
       'nil)
```

11 Sure, using atom and if-true, because '() is an atom.

```
(equal (memb?
         '())
       'nil)
```

Now we can use memb?.

```
(equal (memb? '())
       'nil)
```

12 Once again, xs is '().

```
(equal (if (atom '())
           'nil
           (if (equal (car '()) '?)
               't
               (memb? (cdr '()))))
       'nil)
```

Here is (atom '()) again.

```
(equal (if (atom '())
           'nil
           (if (equal (car '()) '?)
               't
               (memb? (cdr '()))))
       'nil)
```

13 Yes, and after a few simple rewrites this proof is done.

```
(equal 'nil
       'nil)
```

Is memb?/remb0 a theorem?

14 Yes.

How did we prove it?

15 By atom, equal-same, if-true, memb? and remb.

When did we use memb? and remb?

16 When their arguments were '().

Were their arguments always '()?

17 No,
in frame 10 the argument of memb? was an if expression.

How did we reduce the argument of memb? to '()?

18 By using axioms and theorems to simplify ifs and function applications.

What would have happened if we had used memb? before simplifying its argument?

19 The entire if in its argument would be duplicated for each occurrence of xs in the definition of memb?.

What would that have done to the proof of memb?/remb0?

20 We might have had to simplify the duplicated if expression three times instead of once.

Insight: Rewrite from the Inside Out

Rewrite an expression from the "inside" out, starting inside if answers, if elses, and function arguments. Simplify the arguments of a function application as much as possible, then use the Law of Defun to replace the application with the function's body. Rewrite if questions as necessary to use theorems that require premises. Proceed to outer expressions when inner expressions cannot be simplified.

Is memb?/remb1 a theorem?

```
(dethm memb?/remb1 (x1)
  (equal (memb?
           (remb (cons x1 '())))
         'nil))
```

21 We shall see.

As with memb?/remb0, use remb first.

```
(equal (memb?
         (remb (cons x1 '())))
       'nil)
```

22 That claim grew a lot!

```
(equal (memb?
         (if (atom (cons x1 '()))
             '()
             (if (equal (car (cons x1 '())) '?)
                 (remb (cdr (cons x1 '())))
                 (cons (car (cons x1 '()))
                   (remb (cdr (cons x1 '())))))))
       'nil)
```

Why did it grow so much?

23 We substitute xs with the much wider (cons x1 '()) and xs appears five times in the body of the definition of remb.

Next?

```
(equal (memb?
         (if (atom (cons x1 '()))
           '()
           (if (equal (car (cons x1 '())) '?)
             (remb (cdr (cons x1 '())))
             (cons (car (cons x1 '()))
               (remb (cdr (cons x1 '())))))))
       'nil)
```

24 The expression (cons x1 '()) is never an atom.

```
(equal (memb?
         (if (equal (car (cons x1 '())) '?)
           (remb (cdr (cons x1 '())))
           (cons (car (cons x1 '()))
             (remb (cdr (cons x1 '()))))))
       'nil)
```

Now what?

```
(equal (memb?
         (if (equal (car (cons x1 '())) '?)
           (remb (cdr (cons x1 '())))
           (cons (car (cons x1 '()))
             (remb (cdr (cons x1 '()))))))
       'nil)
```

25 Piece of cake. We use car/cons twice and cdr/cons twice.

```
(equal (memb?
         (if (equal x1 '?)
           (remb '())
           (cons x1
             (remb '()))))
       'nil)
```

And now?

26 The argument to these two identical function applications of remb is '() and thus cannot be simplified any more. We could use the Law of Defun and rewrite the applications.

```
(equal (memb?
         (if (equal x1 '?)
           (remb '())
           (cons x1 (remb '()))))
       'nil)
```

Good point. But do we want to rewrite these applications?

27 Why not?

Have we seen the application (remb '()) before?

28 Yes,
it shows up in the theorem memb?/remb0 in frame 9.

Is that significant?

29 Perhaps we can use memb?/remb0 to prove memb?/remb1.

The theorem memb?/remb0 rewrites (memb? (remb '())) to 'nil.

30 In that case, we should not remove the two occurrences of (remb '()) from our expression.

And?

31 We should try to rewrite the expression so that (remb '()) is the argument of memb?.

Exactly! How can we do that?

32 Perhaps we can rewrite the if.

```
(equal (memb?
          (if (equal x1 '?)
              (remb '())
              (cons x1 (remb '()))))
       'nil)
```

Can we use if-true or if-false?

33 No,
we do not know whether (equal x1 '?) is true or false.

Can we use if-same?

34 No,
we cannot rewrite (remb '()) to (cons x1 (remb '())).

If we do not know whether the if's question is true, and the answer and else are not the same, how can we rewrite the if?

35 We do not know. Is there a way?

There is. Recall the rewriting steps in frames 66–71 of chapter 2. First, we have to create a new if using if-same.

```
(equal (memb?
          (if (equal x1 '?)
              (remb '())
              (cons x1 (remb '()))))
       'nil)
```

36 This is certainly familiar.

```
(equal (if (equal x1 '?)
           (memb?
             (if (equal x1 '?)
                 (remb '())
                 (cons x1 (remb '()))))
           (memb?
             (if (equal x1 '?)
                 (remb '())
                 (cons x1 (remb '())))))
       'nil)
```

Can we simplify the argument of memb? using the new if question?

```
(equal (if (equal x1 '?)
           (memb?
             (if (equal x1 '?)
                 (remb '())
                 (cons x1 (remb '()))))
           (memb?
             (if (equal x1 '?)
                 (remb '())
                 (cons x1 (remb '())))))
       'nil)
```

37 We can. In the new if answer, we know that x1 is equal to '?. We use if-nest-A to drop the if in this focus.

```
(equal (if (equal x1 '?)
           (memb?
             (remb '()))
           (memb?
             (if (equal x1 '?)
                 (remb '())
                 (cons x1 (remb '())))))
       'nil)
```

Are we done using the new if?

```
(equal (if (equal x1 '?)
           (memb? (remb '()))
           (memb?
             (if (equal x1 '?)
                 (remb '())
                 (cons x1 (remb '())))))
       'nil)
```

38 In the new if else, x1 is *not* equal to '?. We use if-nest-E to drop another if.

```
(equal (if (equal x1 '?)
           (memb? (remb '()))
           (memb?
             (cons x1 (remb '()))))
       'nil)
```

Here is the familiar expression from memb?/remb0 that we anticipated in frame 31.

```
(equal (if (equal x1 '?)
           (memb? (remb '()))
           (memb?
             (cons x1 (remb '()))))
       'nil)
```

39 Rewriting the entire application of memb? in a new if helps a lot! And now the if question (equal x1 '?) is entirely outside the application.

If Lifting

To move an if question from inside a focus to outside the focus, use if-same where x is the if question and y is the entire focus. This copies the focus in the answer and else of the new if.

```
(original-context                (original-context
  (original-focus          =       (if Q
    (if Q A E)))                     (original-focus
                                       (if Q A E))
                                     (original-focus
                                       (if Q A E))))
```

Then use if-nest-A and if-nest-E to remove each if with the same question in the answer and else of the new if.

```
(original-context
  (if Q                          (original-context
    (original-focus                (if Q
      (if Q A E))          =         (original-focus A)
    (original-focus                  (original-focus E)))
      (if Q A E))))
```

Indeed. It is often useful to move an if question in this way.

40 Good to know.

Insight: Pull Ifs Outward

Use If Lifting when an if is found in an argument of a function application or in an if question. Lift the if outside any function applications and if questions.

Now we can use memb?/remb0.

```
(equal (if (equal x1 '?)
           (memb? (remb '()))
           (memb?
             (cons x1 (remb '()))))
       'nil)
```

41 Yes, we know that this focus is equal to 'nil.

```
(equal (if (equal x1 '?)
           'nil
           (memb?
             (cons x1 (remb '()))))
       'nil)
```

What can we do with the if else?

```
(equal (if (equal x1 '?)
           'nil
           (memb?
             (cons x1 (remb '()))))
       'nil)
```

42 Use memb?.

```
(equal (if (equal x1 '?)
           'nil
           (if (atom (cons x1 (remb '())))
               'nil
               (if (equal (car (cons x1 (remb '())))
                          '?)
                   't
                   (memb?
                     (cdr (cons x1 (remb '())))))))
       'nil)
```

But could we use remb instead, with a different focus?

We could use remb on (remb '()). Do we know anything interesting about (remb '())?

43 As in frame 31, (remb '()) lets us use memb?/remb0, if we can rewrite the expression to form (memb? (remb '())).

It is useful to recognize parts of existing theorems.

44 That much seems clear.

Insight: Keep Theorems in Mind

Bear existing theorems in mind, especially axioms. When the current claim contains an expression that some theorem can rewrite, try using that theorem. When the current claim contains *part* of an expression that some theorem can rewrite, leave that part alone and try to rewrite the current claim in order to use the theorem.

Exactly. Now let's rewrite this focus.

```
(equal (if (equal x1 '?)
           'nil
           (if (atom (cons x1 (remb '())))
               'nil
               (if (equal (car (cons x1 (remb '())))
                          '?)
                   't
                   (memb?
                     (cdr (cons x1 (remb '())))))))
       'nil)
```

45 This is not difficult.

```
(equal (if (equal x1 '?)
           'nil
           (if (equal (car (cons x1 (remb '())))
                      '?)
               't
               (memb?
                 (cdr (cons x1 (remb '()))))))
       'nil)
```

These two are easy.

```
(equal (if (equal x1 '?)
           'nil
           (if (equal (car (cons x1 (remb '())))
                      '?)
               't
               (memb?
                 (cdr (cons x1 (remb '()))))))
       'nil)
```

46 We know car/cons and cdr/cons.

```
(equal (if (equal x1 '?)
           'nil
           (if (equal x1
                      '?)
               't
               (memb?
                 (remb '()))))
       'nil)
```

Does this focus and the orange if question suggest an axiom?

```
(equal (if (equal x1 '?)
           'nil
           (if (equal x1 '?)
               't
               (memb?
                 (remb '()))))
       'nil)
```

47 Yes, they suggest if-nest-E.

```
(equal (if (equal x1 '?)
           'nil
           (memb?
             (remb '())))
       'nil)
```

How about this focus?

```
(equal (if (equal x1 '?)
           'nil
           (memb?
             (remb '())))
       'nil)
```

48 We use memb?/remb0 again, and the proof is finished.

```
(equal (if (equal x1 '?)
           'nil
           'nil)
       'nil)
```

Is memb?/remb1 a theorem?

49 Yes, we have proved it.

What does the theorem memb?/remb1 mean?

50 It means that remb removes '? from a 1-element list.

How did we prove it?

51 By atom/cons, if-false, car/cons, cdr/cons, if-same, if-nest-A, if-nest-E, memb?/remb0, memb?, and remb.

Is this remarkable?

52 Without a doubt.

Let's prove memb?/remb2.

```
(dethm memb?/remb2 (x1 x2)
  (equal (memb?
           (remb
             (cons x2
               (cons x1 '()))))
         'nil))
```

53 This is becoming familiar.

Use remb.

```
(equal (memb?
         (remb (cons x2 (cons x1 '()))))
       'nil)
```

54 That's simple enough.

```
(equal (memb?
         (if (atom (cons x2 (cons x1 '())))
           '()
           (if (equal (car
                        (cons x2 (cons x1 '())))
                      '?)
             (remb
               (cdr (cons x2 (cons x1 '()))))
             (cons
               (car (cons x2 (cons x1 '())))
               (remb
                 (cdr
                   (cons x2 (cons x1 '()))))))))
       'nil)
```

Do we know the if question (atom (cons x2 (cons x1 '())))?

```
(equal (memb?
         (if (atom (cons x2 (cons x1 '())))
           '()
           (if (equal (car (cons x2 (cons x1 '())))
                      '?)
             (remb
               (cdr (cons x2 (cons x1 '()))))
             (cons
               (car (cons x2 (cons x1 '())))
               (remb
                 (cdr
                   (cons x2 (cons x1 '()))))))))
       'nil)
```

55 'nil.

```
(equal (memb?
         (if (equal (car (cons x2 (cons x1 '())))
                    '?)
           (remb
             (cdr (cons x2 (cons x1 '()))))
           (cons
             (car (cons x2 (cons x1 '())))
             (remb
               (cdr
                 (cons x2 (cons x1 '())))))))
       'nil)
```

Deal with the cars and cdrs.

```
(equal (memb?
         (if (equal (car (cons x2 (cons x1 '())))
                    '?)
           (remb
             (cdr (cons x2 (cons x1 '()))))
           (cons
             (car (cons x2 (cons x1 '())))
             (remb
               (cdr
                 (cons x2 (cons x1 '())))))))
       'nil)
```

56 Done.

```
(equal (memb?
         (if (equal x2 '?)
           (remb (cons x1 '()))
           (cons x2
             (remb (cons x1 '())))))
       'nil)
```

In this case . . .

```
(equal (memb?
         (if (equal x2 '?)
           (remb (cons x1 '()))
           (cons x2
             (remb (cons x1 '())))))
       'nil)
```

57 We use If Lifting!

```
(equal (if (equal x2 '?)
         (memb?
           (remb (cons x1 '())))
         (memb?
           (cons x2
             (remb (cons x1 '())))))
       'nil)
```

Exactly. Which axioms do we use in frame 57?

58 We use if-same, if-nest-A, and if-nest-E, as in frames 36–38.

Are we experiencing déjà vu?

```
(equal (if (equal x2 '?)
         (memb?
           (remb (cons x1 '())))
         (memb?
           (cons x2
             (remb (cons x1 '())))))
       'nil)
```

59 Yes, we saw this expression in the theorem memb?/remb1. We can use memb?/remb1 here.

```
(equal (if (equal x2 '?)
         'nil
         (memb?
           (cons x2
             (remb (cons x1 '())))))
       'nil)
```

Let's use memb?. Ready?

```
(equal (if (equal x2 '?)
           'nil
           (memb?
             (cons x2
               (remb (cons x1 '())))))
       'nil)
```

60 And willing.

```
(equal (if (equal x2 '?)
           'nil
           (if (atom
                 (cons x2
                   (remb (cons x1 '()))))
              'nil
              (if (equal
                    (car
                      (cons x2
                        (remb (cons x1 '()))))
                    '?)
                 't
                 (memb?
                   (cdr
                     (cons x2
                       (remb (cons x1 '())))))))))
       'nil)
```

The next two steps are easy.

```
(equal (if (equal x2 '?)
           'nil
           (if (atom
                 (cons x2
                   (remb (cons x1 '()))))
              'nil
              (if (equal (car
                           (cons x2
                             (remb (cons x1 '()))))
                         '?)
                 't
                 (memb?
                   (cdr
                     (cons x2
                       (remb (cons x1 '())))))))))
       'nil)
```

61 That's music to our ears.

```
(equal (if (equal x2 '?)
           'nil
           (if (equal (car
                        (cons x2
                          (remb (cons x1 '()))))
                      '?)
              't
              (memb?
                (cdr
                  (cons x2
                    (remb (cons x1 '())))))))
       'nil)
```

So are these two steps.

```
(equal (if (equal x2 '?)
           'nil
           (if (equal (car
                        (cons x2
                          (remb (cons x1 '()))))
                      '?)
               't
               (memb?
                 (cdr
                   (cons x2
                     (remb (cons x1 '())))))))
       'nil)
```

62 Those steps sure are easy.

```
(equal (if (equal x2 '?)
           'nil
           (if (equal x2 '?)
               't
               (memb?
                 (remb (cons x1 '())))))
       'nil)
```

Do we know (equal x2 '?) in this focus?

```
(equal (if (equal x2 '?)
           'nil
           (if (equal x2 '?)
               't
               (memb? (remb (cons x1 '())))))
       'nil)
```

63 Yes, by if-nest-E.

```
(equal (if (equal x2 '?)
           'nil
           (memb? (remb (cons x1 '()))))
       'nil)
```

It's déjà vu all over again![†]

```
(equal (if (equal x2 '?)
           'nil
           (memb? (remb (cons x1 '()))))
       'nil)
```

[†]Yogi Berra might have said so.

64 It's quite convenient that memb?/remb1 is a theorem.

```
(equal (if (equal x2 '?)
           'nil
           'nil)
       'nil)
```

Is memb?/remb2 a theorem?

65 Certainly.

What does memb?/remb2 mean?

66 The same as memb?/remb0 and memb?/remb1, but for 2-element lists.

How does the proof of memb?/remb2 compare to the proof of memb?/remb1?

67 It uses the same steps in the same order, but with slightly longer lists and using memb?/remb1 instead of memb?/remb0.

Can we prove memb?/remb3?

```
(dethm memb?/remb3 (x1 x2 x3)
  (equal
    (memb?
      (remb
        (cons x3
          (cons x2
            (cons x1 '())))))
    'nil))
```

68 Sure,
just use the same approach and rely on memb?/remb2.

Of course.

69 How many times must we do this?

Until we reach the largest list.

70 That will take forever and ever.

It certainly will.

71 And we are getting hungry!

Take a short visit to J-Bob, first. The proofs about memb? and remb start on page 185.

72 Then something tasty?

Why not take a break and have a steak.

73 That is indeed a rare breakfast.

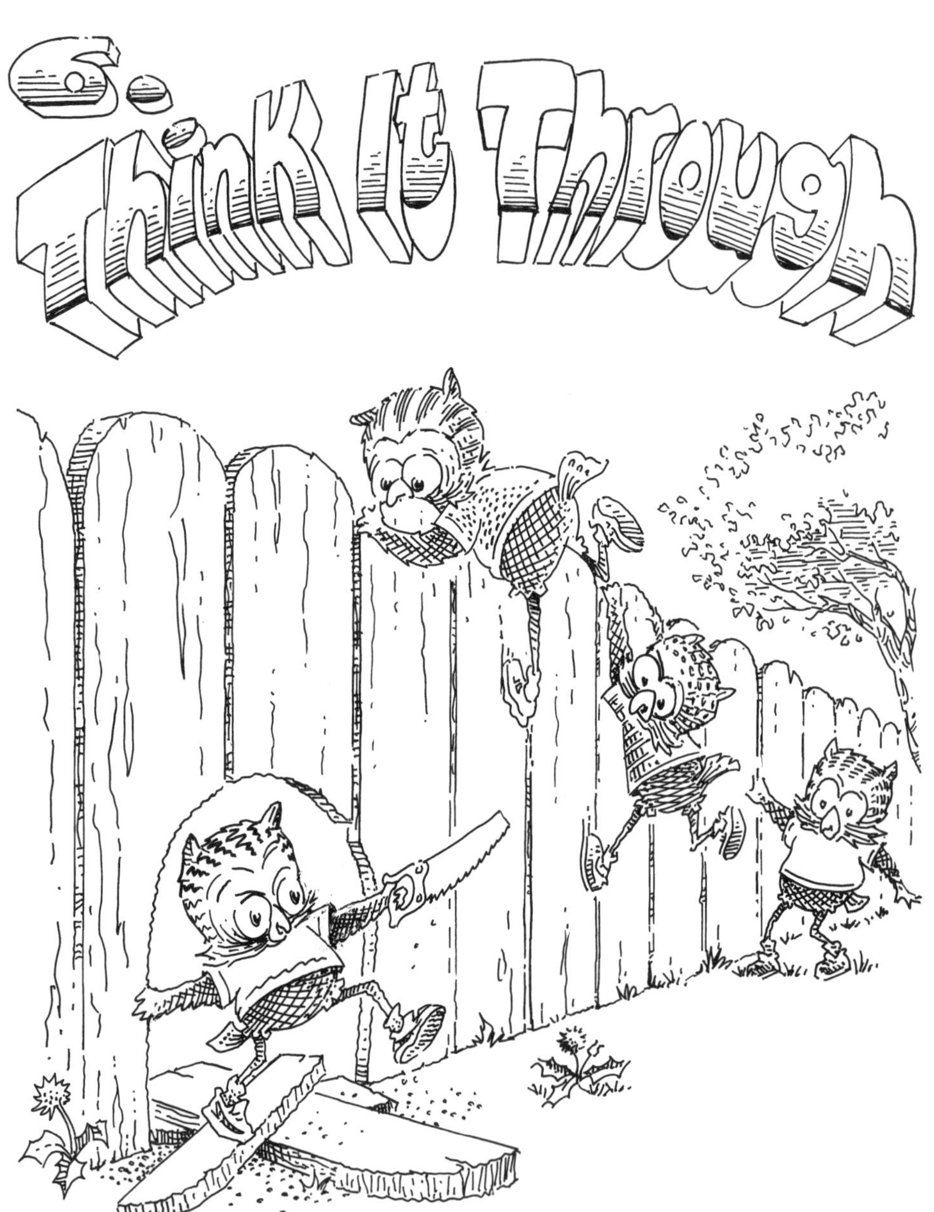
6. Think It Through

Are the proofs of memb?/remb0, memb?/remb1, and memb?/remb2 building toward a larger theorem?

1 It seems so.

What should the theorem say?

2 It should say that remb removes '? from any list.

Going one element at a time isn't working well enough.

3 This is just like list0?, list1?, and list2? in chapter 4.

How so?

4 We start with the empty list and define a new function for each slightly longer list.

How do those new functions work?

5 They deal with one element, then call the previous function.

How do we extend a function to work on lists of any length?

6 The answer is recursion!

We like simple answers.

7 Can we use recursion in proofs?

Yes,
recursion in proofs is called *induction* and it's a great idea. How do we think that would work?

8 The same way as in functions. One case for the empty list and handle the rest via *natural recursion.*

And what is the natural recursion?

9 For a list xs, it is the same function call but with (cdr xs).

Shall we try out this idea?

10 We can't wait!

Try to prove memb?/remb.

```
(dethm memb?/remb (xs)
  (equal (memb? (remb xs)) 'nil))
```

11 Here we go again.

But only one more time. Here is the inductive claim we must prove for memb?/remb.

```
(if (atom xs)
  (equal (memb? (remb xs)) 'nil)
  (if (equal (memb? (remb (cdr xs))) 'nil)
    (equal (memb? (remb xs)) 'nil)
    't))
```

Which parts of the if answer and if else are identical?

12 The parts that are exactly the same as the claim memb?/remb.

```
(if (atom xs)
  (equal (memb? (remb xs)) 'nil)
  (if (equal (memb? (remb (cdr xs))) 'nil)
    (equal (memb? (remb xs)) 'nil)
    't))
```

The first if answer states memb?/remb for empty lists, where xs is an atom. What about the other if answer?

```
(if (atom xs)
  (equal (memb? (remb xs)) 'nil)
  (if (equal (memb? (remb (cdr xs))) 'nil)
    (equal (memb? (remb xs)) 'nil)
    't))
```

13 This one states memb?/remb for non-empty lists, where xs has at least one cons.

```
(if (atom xs)
  (equal (memb? (remb xs)) 'nil)
  (if (equal (memb? (remb (cdr xs))) 'nil)
    (equal (memb? (remb xs)) 'nil)
    't))
```

The orange expression is not quite the same as memb?/remb. How is it different?

```
(if (atom xs)
  (equal (memb? (remb xs)) 'nil)
  (if (equal (memb? (remb (cdr xs))) 'nil)
    (equal (memb? (remb xs)) 'nil)
    't))
```

14 This expression has (cdr xs) where memb?/remb has xs. It looks like the natural recursion for our claim.

The (natural) recursion for a claim is called the *inductive premise*. What is it for?

15 We're not sure. Once we use it, perhaps we'll find out.

How does our proof begin?

```
(if (atom xs)
  (equal (memb?
           (remb xs))
         'nil)
  (if (equal (memb? (remb (cdr xs))) 'nil)
    (equal (memb? (remb xs)) 'nil)
    't))
```

16 We use the definition of remb.

```
(if (atom xs)
  (equal (memb?
           (if (atom xs)
             '()
             (if (equal (car xs) '?)
               (remb (cdr xs))
               (cons (car xs)
                 (remb (cdr xs))))))
         'nil)
  (if (equal (memb? (remb (cdr xs))) 'nil)
    (equal (memb? (remb xs)) 'nil)
    't))
```

We know which way this if goes, based on the premise (atom xs).

```
(if (atom xs)
  (equal (memb?
           (if (atom xs)
             '()
             (if (equal (car xs) '?)
               (remb (cdr xs))
               (cons (car xs)
                 (remb (cdr xs))))))
         'nil)
  (if (equal (memb? (remb (cdr xs))) 'nil)
    (equal (memb? (remb xs)) 'nil)
    't))
```

17 Yes, that nested if is obvious.

```
(if (atom xs)
  (equal (memb?
           '())
         'nil)
  (if (equal (memb? (remb (cdr xs))) 'nil)
    (equal (memb? (remb xs)) 'nil)
    't))
```

Now?

```
(if (atom xs)
  (equal (memb? '())
         'nil)
  (if (equal (memb? (remb (cdr xs))) 'nil)
    (equal (memb? (remb xs)) 'nil)
    't))
```

18 Use memb?. Easy.

```
(if (atom xs)
  (equal (if (atom '())
             'nil
             (if (equal (car '()) '?)
               't
               (memb? (cdr '()))))
         'nil)
  (if (equal (memb? (remb (cdr xs))) 'nil)
    (equal (memb? (remb xs)) 'nil)
    't))
```

Can we simplify this answer?

```
(if (atom xs)
  (equal (if (atom '())
             'nil
             (if (equal (car '()) '?)
               't
               (memb? (cdr '()))))
         'nil)
  (if (equal (memb? (remb (cdr xs))) 'nil)
    (equal (memb? (remb xs)) 'nil)
    't))
```

19 Yes, using atom, if-true, and equal-same.

```
(if (atom xs)
  't
  (if (equal (memb? (remb (cdr xs))) 'nil)
    (equal (memb? (remb xs)) 'nil)
    't))
```

Consider the case for non-empty lists.

```
(if (atom xs)
  't
  (if (equal (memb? (remb (cdr xs))) 'nil)
    (equal (memb?
             (remb xs))
           'nil)
    't))
```

20 We start by using remb.

```
(if (atom xs)
  't
  (if (equal (memb? (remb (cdr xs))) 'nil)
    (equal (memb?
             (if (atom xs)
               '()
               (if (equal (car xs) '?)
                 (remb (cdr xs))
                 (cons (car xs)
                   (remb (cdr xs))))))
           'nil)
    't))
```

Remember, xs is non-empty in this focus since it is in the if else of the premise (atom xs).

```
(if (atom xs)
  't
  (if (equal (memb? (remb (cdr xs))) 'nil)
    (equal (memb?
              (if (atom xs)
                '()
                (if (equal (car xs) '?)
                  (remb (cdr xs))
                  (cons (car xs)
                    (remb (cdr xs))))))
           'nil)
    't))
```

21 We can use if-nest-E.

```
(if (atom xs)
  't
  (if (equal (memb? (remb (cdr xs))) 'nil)
    (equal (memb?
              (if (equal (car xs) '?)
                (remb (cdr xs))
                (cons (car xs)
                  (remb (cdr xs)))))
           'nil)
    't))
```

What do we do here?

```
(if (atom xs)
  't
  (if (equal (memb? (remb (cdr xs))) 'nil)
    (equal (memb?
              (if (equal (car xs) '?)
                (remb (cdr xs))
                (cons (car xs)
                  (remb (cdr xs)))))
           'nil)
    't))
```

22 Like memb?/remb1 and memb?/remb2, we use If Lifting on (equal (car xs) '?).

```
(if (atom xs)
  't
  (if (equal (memb? (remb (cdr xs))) 'nil)
    (equal (if (equal (car xs) '?)
             (memb? (remb (cdr xs)))
             (memb?
               (cons (car xs)
                 (remb (cdr xs)))))
           'nil)
    't))
```

Here, this amounts to pushing memb? into the answer and else of an if.

The inductive premise states that this focus is equal to 'nil. Can we do anything with it?

```
(if (atom xs)
  't
  (if (equal (memb? (remb (cdr xs))) 'nil)
    (equal (if (equal (car xs) '?)
               (memb? (remb (cdr xs)))
               (memb?
                 (cons (car xs)
                   (remb (cdr xs)))))
           'nil)
    't))
```

23 By equal-if, we are able to replace this focus with 'nil.

```
(if (atom xs)
  't
  (if (equal (memb? (remb (cdr xs))) 'nil)
    (equal (if (equal (car xs) '?)
               'nil
               (memb?
                 (cons (car xs)
                   (remb (cdr xs)))))
           'nil)
    't))
```

What next?

24 Perhaps we could use remb here.

```
(if (atom xs)
  't
  (if (equal (memb? (remb (cdr xs))) 'nil)
    (equal (if (equal (car xs) '?)
               'nil
               (memb?
                 (cons (car xs)
                   (remb (cdr xs)))))
           'nil)
    't))
```

What do we know about (remb (cdr xs))?

25 According to the inductive premise, we know that (memb? (remb (cdr xs))) is equal to 'nil. But the inductive premise does not allow us to rewrite (remb (cdr xs)) itself.

Just like with existing theorems, always keep the inductive premise in mind. Never rewrite the natural recursion in an inductive proof; keep it around until the inductive premise helps simplify it.

26 We'll remember that.

Insight: Don't Touch Inductive Premises

Do not try to simplify an inductive premise in an inductive proof directly. Instead, rewrite the expression around it until the inductive premise can be applied. Often, after applying the inductive premise, an inductive proof is nearly done.

Use memb? in this focus.

```
(if (atom xs)
  't
  (if (equal (memb? (remb (cdr xs))) 'nil)
    (equal (if (equal (car xs) '?)
                'nil
                (memb?
                  (cons (car xs)
                    (remb (cdr xs)))))
              'nil)
    't))
```

27 That's easy.

```
(if (atom xs)
  't
  (if (equal (memb? (remb (cdr xs))) 'nil)
    (equal (if (equal (car xs) '?)
                'nil
                (if (atom (cons (car xs)
                                (remb (cdr xs))))
                  'nil
                  (if (equal (car
                               (cons (car xs)
                                 (remb (cdr xs))))
                            '?)
                    't
                    (memb?
                      (cdr
                        (cons (car xs)
                          (remb (cdr xs))))))))
              'nil)
    't))
```

Next?

```
(if (atom xs)
  't
  (if (equal (memb? (remb (cdr xs))) 'nil)
    (equal (if (equal (car xs) '?)
               'nil
               (if (atom (cons (car xs)
                               (remb (cdr xs))))
                 'nil
                 (if (equal (car
                             (cons (car xs)
                               (remb (cdr xs))))
                           '?)
                   't
                   (memb?
                     (cdr
                       (cons (car xs)
                         (remb
                           (cdr xs))))))))
           'nil)
    't))
```

28 We can drop one if.

```
(if (atom xs)
  't
  (if (equal (memb? (remb (cdr xs))) 'nil)
    (equal (if (equal (car xs) '?)
               'nil
               (if (equal (car
                            (cons (car xs)
                              (remb (cdr xs))))
                          '?)
                 't
                 (memb?
                   (cdr
                     (cons (car xs)
                       (remb
                         (cdr xs)))))))
           'nil)
    't))
```

Here are two easy steps.

```
(if (atom xs)
  't
  (if (equal (memb? (remb (cdr xs))) 'nil)
    (equal (if (equal (car xs) '?)
               'nil
               (if (equal (car
                            (cons (car xs)
                              (remb
                                (cdr xs))))
                          '?)
                 't
                 (memb?
                   (cdr
                     (cons (car xs)
                       (remb
                         (cdr xs)))))))
           'nil)
    't))
```

29 We can use car/cons and cdr/cons in our sleep at this point.

```
(if (atom xs)
  't
  (if (equal (memb? (remb (cdr xs))) 'nil)
    (equal (if (equal (car xs) '?)
               'nil
               (if (equal (car xs)
                          '?)
                 't
                 (memb?
                   (remb
                     (cdr xs)))))
           'nil)
    't))
```

Anything special about the if in this focus?

```
(if (atom xs)
  't
  (if (equal (memb? (remb (cdr xs))) 'nil)
    (equal (if (equal (car xs) '?)
               'nil
               (if (equal (car xs) '?)
                 't
                 (memb?
                   (remb (cdr xs)))))
           'nil)
    't))
```

30 It is nested, of course.

```
(if (atom xs)
  't
  (if (equal (memb? (remb (cdr xs))) 'nil)
    (equal (if (equal (car xs) '?)
               'nil
               (memb?
                 (remb (cdr xs))))
           'nil)
    't))
```

What do we see in this focus?

```
(if (atom xs)
  't
  (if (equal (memb? (remb (cdr xs))) 'nil)
    (equal (if (equal (car xs) '?)
               'nil
               (memb? (remb (cdr xs))))
           'nil)
    't))
```

31 Another chance to use the inductive premise.

```
(if (atom xs)
  't
  (if (equal (memb? (remb (cdr xs))) 'nil)
    (equal (if (equal (car xs) '?)
               'nil
               'nil)
           'nil)
    't))
```

This is rather obvious.

```
(if (atom xs)
  't
  (if (equal (memb? (remb (cdr xs))) 'nil)
    (equal (if (equal (car xs) '?)
               'nil
               'nil)
           'nil)
    't))
```

32 Certainly.

```
(if (atom xs)
  't
  (if (equal (memb? (remb (cdr xs))) 'nil)
    't
    't))
```

And now the proof of memb?/remb is done.

33 Q.E.D.

Does the proof of memb?/remb look familiar?

34 Yes, it looks like the proofs we did for memb?/remb0, memb?/remb1, and memb?/remb2.

Insight: Build Up to Induction Gradually

Build up to a proof by induction over lists by proving theorems about the empty list, lists with one element, lists with two elements, and so on. Once the pattern of these proofs is clear, the proof by induction should be similar.

---

35 Congratulations! We have learned how to prove a theorem with induction.

Can we use induction for other proofs?

Proof by List Induction

To prove a claim C by induction over a list named x, prove

(if (atom x) C (if C_{cdr} C 't))

where C_{cdr} is C with x replaced by (cdr x).

---

36 Have we learned the answer to the ultimate question of life, the universe, and everything yet?†

†Thank you, Douglas Adams (1952-2001).

No, but we have learned how to do induction.

---

37 Does that mean we know what the inductive premise is for?

No, not completely. Perhaps, reading this chapter one more time would help?

---

38 Perhaps. J-Bob can help with induction; the proof of memb?/remb starts on page 187.

Will do.

---

39 It is time for a hot fudge sundae.

Not exactly a breakfast of champions.

7
Oh My, Stars!

1 The function ctx? determines whether its argument contains '?. What value is (ctx? '()) equal to?

'nil, because there is no '? in '().

2 What value is (ctx? '(a (? ?) c)) equal to?

't, because there is a '? in '(a (? ?) c).

3 Define ctx?.

```
(defun ctx? (x)
  (if (atom x)
      (equal x '?)
      (if (ctx? (car x))
          't
          (ctx? (cdr x)))))

measure: (size x)
```

4 We can show that ctx? is total. Here is the claim we must prove.†

```
(if (natp (size x))
    (if (atom x)
        't
        (if (< (size (car x)) (size x))
            (if (ctx? (car x))
                't
                (< (size (cdr x)) (size x)))
            '()))
    '())
```

Although ctx? may not be total, we can apply ctx? in its totality claim so long as we do not use its definition via the Law of Defun.

†This totality claim is more complex than most others in this book. In chapter 8, we describe the process of creating totality claims in detail.

Using natp/size, we know that (size x) is a natural number. We can drop an if.

```
(if (atom x)
    't
    (if (< (size (car x)) (size x))
        (if (ctx? (car x))
            't
            (< (size (cdr x)) (size x)))
        '()))
```

What can we do with the size comparisons using this premise?

```
(if (atom x)
  't
  (if (< (size (car x)) (size x))
    (if (ctx? (car x))
      't
      (< (size (cdr x)) (size x)))
    '()))
```

5 We can use size/car on the first application of < and size/cdr on the second application of <.

```
(if (atom x)
  't
  (if 't
    (if (ctx? (car x))
      't
      't)
    '()))
```

And then?

6 The proof is essentially done.

Recall sub from frame 88 of chapter 4.

```
(defun sub (x y)
  (if (atom y)
    (if (equal y '?)
      x
      y)
    (cons (sub x (car y))
      (sub x (cdr y)))))

measure: (size y)
```

State the claim that if x and y contain '?, then so does (sub x y).

7

```
(dethm ctx?/sub (x y)
  (if (ctx? x)
    (if (ctx? y)
      (equal (ctx? (sub x y)) 't)
      't)
    't))
```

Is ctx?/sub a theorem?

We have to prove it to find out. Can we use induction to prove it?

8 Maybe,
but sub and ctx? are not quite like remb and memb?.

How do they differ?

9 The functions sub and ctx? perform natural recursion on the car, as well as the cdr. The functions remb and memb? only perform natural recursion on the cdr.

Proof by Star Induction

To prove a claim C by induction over cars and cdrs of a variable named x, prove

(if (atom x) C (if C_{car} (if C_{cdr} C 't) 't))

where C_{car} is C with x replaced by (car x) and C_{cdr} is C with x replaced by (cdr x).

Use our new kind of induction over y.

```
(if (ctx? x)
  (if (ctx? y)
    (equal (ctx? (sub x y)) 't)
    't)
  't)
```

10 Wow, using induction enlarges the claim!

```
(if (atom y)
  (if (ctx? x)
    (if (ctx? y)
      (equal (ctx? (sub x y)) 't)
      't)
    't)
  (if (if (ctx? x)
        (if (ctx? (car y))
          (equal (ctx? (sub x (car y))) 't)
          't)
        't)
    (if (if (ctx? x)
          (if (ctx? (cdr y))
            (equal (ctx? (sub x (cdr y))) 't)
            't)
          't)
      (if (ctx? x)
        (if (ctx? y)
          (equal (ctx? (sub x y)) 't)
          't)
        't)
      't)
    't))
```

We now have an inductive premise for the car as well as the cdr. There are four versions of our original claim now, written in orange. One is the case where y is an atom, one is the case where y is a cons, and the other two are inductive premises. Can we make this claim smaller?

```
(if (atom y)
  (if (ctx? x)
    (if (ctx? y)
      (equal (ctx? (sub x y)) 't)
      't)
    't)
  (if (if (ctx? x)
        (if (ctx? (car y))
          (equal (ctx? (sub x (car y))) 't)
          't)
        't)
    (if (if (ctx? x)
          (if (ctx? (cdr y))
            (equal (ctx? (sub x (cdr y))) 't)
            't)
          't)
      (if (ctx? x)
        (if (ctx? y)
          (equal (ctx? (sub x y)) 't)
          't)
        't)
      't)
    't))
```

11 Yes, we do so by using If Lifting on (ctx? x). We also use if-same thrice to simplify the final if else to 't.

```
(if (ctx? x)
  (if (atom y)
    (if (ctx? y)
      (equal (ctx? (sub x y)) 't)
      't)
    (if (if (ctx? (car y))
          (equal (ctx? (sub x (car y))) 't)
          't)
      (if (if (ctx? (cdr y))
            (equal (ctx? (sub x (cdr y))) 't)
            't)
        (if (ctx? y)
          (equal (ctx? (sub x y)) 't)
          't)
        't)
      't))
  't)
```

Insight: Combine Ifs

When there are multiple ifs with the same question, combine them into one if with If Lifting. Lift the ifs outside any function applications and if questions.

Next, use sub.

```
(if (ctx? x)
  (if (atom y)
    (if (ctx? y)
      (equal (ctx? (sub x y))
             't)
      't)
    (if (if (ctx? (car y))
          (equal (ctx? (sub x (car y))) 't)
          't)
      (if (if (ctx? (cdr y))
            (equal (ctx? (sub x (cdr y))) 't)
            't)
        (if (ctx? y)
            (equal (ctx? (sub x y)) 't)
            't)
          't)
      't))
  't)
```

12 And if-nest-A using the premise (atom y).

```
(if (ctx? x)
  (if (atom y)
    (if (ctx? y)
      (equal (ctx? (if (equal y '?)
                       x
                       y))
             't)
      't)
    (if (if (ctx? (car y))
          (equal (ctx? (sub x (car y))) 't)
          't)
      (if (if (ctx? (cdr y))
            (equal (ctx? (sub x (cdr y))) 't)
            't)
        (if (ctx? y)
            (equal (ctx? (sub x y)) 't)
            't)
          't)
      't))
  't)
```

Do we want to expand
(ctx? (if (equal y '?) x y))?

```
(if (ctx? x)
  (if (atom y)
    (if (ctx? y)
      (equal (ctx? (if (equal y '?)
                       x
                       y))
             't)
      't)
    (if (if (ctx? (car y))
          (equal (ctx? (sub x (car y))) 't)
          't)
      (if (if (ctx? (cdr y))
            (equal (ctx? (sub x (cdr y))) 't)
            't)
        (if (ctx? y)
            (equal (ctx? (sub x y)) 't)
            't)
          't)
      't))
  't)
```

13 No, we want to lift the if outside the applications of ctx? and equal.

```
(if (ctx? x)
  (if (atom y)
    (if (ctx? y)
      (if (equal y '?)
        (equal (ctx? x) 't)
        (equal (ctx? y) 't))
      't)
    (if (if (ctx? (car y))
          (equal (ctx? (sub x (car y))) 't)
          't)
      (if (if (ctx? (cdr y))
            (equal (ctx? (sub x (cdr y))) 't)
            't)
        (if (ctx? y)
            (equal (ctx? (sub x y)) 't)
            't)
          't)
      't))
  't)
```

14 If the premise (ctx? x) is true, can we rewrite (ctx? x) to 't in this if answer?

```
(if (ctx? x)
   (if (atom y)
      (if (ctx? y)
         (if (equal y '?)
            (equal (ctx? x) 't)
            (equal (ctx? y) 't))
         't)
      (if (if (ctx? (car y))
             (equal (ctx? (sub x (car y))) 't)
             't)
         (if (if (ctx? (cdr y))
                (equal (ctx? (sub x (cdr y))) 't)
                't)
            (if (ctx? y)
               (equal (ctx? (sub x y)) 't)
               't)
            't)
         't))
   't)
```

That sounds right. Shouldn't we be able to rewrite any premise to 't if it is true?

---

15 Not necessarily, since if is total. Its question might have a value other than 't or 'nil.

That's true, and also surprising.

---

16 So, can we rewrite (ctx? x) to 't when it is true?

We should be able to, since ctx? only produces 't or 'nil.

---

17 What theorem rewrites (ctx? x) to 't?

We do not know of any. We must first prove that (ctx? x) is equal to 't.

---

18 How many steps does that take?

We don't know. The function ctx? is recursive, so we may need to use induction again.

In that case, let's state a new claim that if (ctx? x) is true, it is equal to 't.

19 That is easy.

```
(dethm ctx?/t (x)
  (if (ctx? x)
      (equal (ctx? x) 't)
      't))
```

Insight: Create Helpers for Induction

To rewrite the application of a recursive function, prove a separate theorem about the recursive function using induction. Do this if the current proof either does not use induction, or uses induction for a different kind of recursion from the function, or uses induction on different arguments from the application.

Can we use ctx?/t now?

```
(if (ctx? x)
  (if (atom y)
    (if (ctx? y)
      (if (equal y '?)
        (equal (ctx? x) 't)
        (equal (ctx? y) 't))
      't)
    (if (if (ctx? (car y))
          (equal (ctx? (sub x (car y))) 't)
          't)
      (if (if (ctx? (cdr y))
            (equal (ctx? (sub x (cdr y))) 't)
            't)
        (if (ctx? y)
          (equal (ctx? (sub x y)) 't)
          't)
        't)
      't))
  't)
```

20 Yes, using these premises, if we prove ctx?/t eventually.

```
(if (ctx? x)
  (if (atom y)
    (if (ctx? y)
      (if (equal y '?)
        (equal 't 't)
        (equal 't 't))
      't)
    (if (if (ctx? (car y))
          (equal (ctx? (sub x (car y))) 't)
          't)
      (if (if (ctx? (cdr y))
            (equal (ctx? (sub x (cdr y))) 't)
            't)
        (if (ctx? y)
          (equal (ctx? (sub x y)) 't)
          't)
        't)
      't))
  't)
```

What can we do with this focus?

```
(if (ctx? x)
  (if (atom y)
    (if (ctx? y)
      (if (equal y '?)
        (equal 't 't)
        (equal 't 't))
      't)
    (if (if (ctx? (car y))
          (equal (ctx? (sub x (car y))) 't)
          't)
      (if (if (ctx? (cdr y))
            (equal (ctx? (sub x (cdr y))) 't)
            't)
        (if (ctx? y)
          (equal (ctx? (sub x y)) 't)
          't)
        't)
      't))
  't)
```

21 We use equal-same and if-same to rewrite the entire focus to 't.

```
(if (ctx? x)
  (if (atom y)
    't
    (if (if (ctx? (car y))
          (equal (ctx? (sub x (car y))) 't)
          't)
      (if (if (ctx? (cdr y))
            (equal (ctx? (sub x (cdr y))) 't)
            't)
        (if (ctx? y)
          (equal (ctx? (sub x y)) 't)
          't)
        't)
      't))
  't)
```

Does ctx?/t need to be proved eventually?

22 Yes, it certainly does.

Use sub and simplify the result.

```
(if (ctx? x)
  (if (atom y)
    't
    (if (if (ctx? (car y))
          (equal (ctx? (sub x (car y))) 't)
          't)
      (if (if (ctx? (cdr y))
            (equal (ctx? (sub x (cdr y))) 't)
            't)
        (if (ctx? y)
          (equal (ctx? (sub x y))
                 't)
          't)
        't)
      't))
  't)
```

23 Very well, we use if-nest-E and the premise (atom y).

```
(if (ctx? x)
  (if (atom y)
    't
    (if (if (ctx? (car y))
          (equal (ctx? (sub x (car y))) 't)
          't)
      (if (if (ctx? (cdr y))
            (equal (ctx? (sub x (cdr y))) 't)
            't)
        (if (ctx? y)
          (equal (ctx? (cons (sub x (car y))
                             (sub x (cdr y))))
                 't)
          't)
        't)
      't))
  't)
```

Now use ctx?. Again, simplify the result.

```
(if (ctx? x)
  (if (atom y)
    't
    (if (if (ctx? (car y))
            (equal (ctx? (sub x (car y))) 't)
            't)
        (if (if (ctx? (cdr y))
                (equal (ctx? (sub x (cdr y))) 't)
                't)
            (if (ctx? y)
                (equal (ctx? (cons (sub x (car y))
                                   (sub x (cdr y))))
                       't)
                't)
            't)
        't))
  't)
```

24 This time we use atom/cons, car/cons, cdr/cons, and if-false.

```
(if (ctx? x)
  (if (atom y)
    't
    (if (if (ctx? (car y))
            (equal (ctx? (sub x (car y))) 't)
            't)
        (if (if (ctx? (cdr y))
                (equal (ctx? (sub x (cdr y))) 't)
                't)
            (if (ctx? y)
                (equal (if (ctx? (sub x (car y)))
                           't
                           (ctx? (sub x (cdr y))))
                       't)
                't)
            't)
        't))
  't)
```

Do we want to rewrite the applications of ctx? written in orange?

```
(if (ctx? x)
  (if (atom y)
    't
    (if (if (ctx? (car y))
            (equal (ctx? (sub x (car y))) 't)
            't)
        (if (if (ctx? (cdr y))
                (equal (ctx? (sub x (cdr y))) 't)
                't)
            (if (ctx? y)
                (equal (if (ctx? (sub x (car y)))
                           't
                           (ctx? (sub x (cdr y))))
                       't)
                't)
            't)
        't))
  't)
```

25 No, they are each part of an inductive premise. We rewrite the if question above them, instead.

```
(if (ctx? x)
  (if (atom y)
    't
    (if (if (ctx? (car y))
            (equal (ctx? (sub x (car y))) 't)
            't)
        (if (if (ctx? (cdr y))
                (equal (ctx? (sub x (cdr y))) 't)
                't)
            (if (if (ctx? (car y))
                    't
                    (ctx? (cdr y)))
                (equal (if (ctx? (sub x (car y)))
                           't
                           (ctx? (sub x (cdr y))))
                       't)
                't)
            't)
        't))
  't)
```

There are two ifs with the same question.

```
(if (ctx? x)
    (if (atom y)
        't
        (if (if (ctx? (car y))
                (equal (ctx? (sub x (car y))) 't)
                't)
            (if (if (ctx? (cdr y))
                    (equal (ctx? (sub x (cdr y))) 't)
                    't)
                (if (if (ctx? (car y))
                        't
                        (ctx? (cdr y)))
                    (equal (if (ctx? (sub x (car y)))
                               't
                               (ctx? (sub x (cdr y))))
                           't)
                    't)
                't)
            't))
    't)
```

26 Clearly, we should lift that if question. We also use if-true twice.

```
(if (ctx? x)
    (if (atom y)
        't
        (if (ctx? (car y))
            (if (equal (ctx? (sub x (car y))) 't)
                (if (if (ctx? (cdr y))
                        (equal (ctx? (sub x (cdr y))) 't)
                        't)
                    (equal (if (ctx? (sub x (car y)))
                               't
                               (ctx? (sub x (cdr y))))
                           't)
                    't)
                't)
            (if (if (ctx? (cdr y))
                    (equal (ctx? (sub x (cdr y))) 't)
                    't)
                (if (ctx? (cdr y))
                    (equal (if (ctx? (sub x (car y)))
                               't
                               (ctx? (sub x (cdr y))))
                           't)
                    't)
                't)))
    't)
```

Now we can use an inductive premise.

```
(if (ctx? x)
    (if (atom y)
        't
        (if (ctx? (car y))
            (if (equal (ctx? (sub x (car y))) 't)
                (if (if (ctx? (cdr y))
                        (equal (ctx? (sub x (cdr y))) 't)
                        't)
                    (equal (if (ctx? (sub x (car y)))
                               't
                               (ctx? (sub x (cdr y))))
                           't)
                    't)
                't)
            (if (if (ctx? (cdr y))
                    (equal (ctx? (sub x (cdr y))) 't)
                    't)
                (if (ctx? (cdr y))
                    (equal (if (ctx? (sub x (car y)))
                               't
                               (ctx? (sub x (cdr y))))
                           't)
                    't)
                't)))
    't)
```

27 And rewrite another if answer to 't.

```
(if (ctx? x)
    (if (atom y)
        't
        (if (ctx? (car y))
            't
            (if (if (ctx? (cdr y))
                    (equal (ctx? (sub x (cdr y))) 't)
                    't)
                (if (ctx? (cdr y))
                    (equal (if (ctx? (sub x (car y)))
                               't
                               (ctx? (sub x (cdr y))))
                           't)
                    't)
                't)))
    't)
```

Once again, there are two ifs with the same question.

```
(if (ctx? x)
    (if (atom y)
        't
        (if (ctx? (car y))
            't
            (if (if (ctx? (cdr y))
                    (equal (ctx? (sub x (cdr y))) 't)
                    't)
                (if (ctx? (cdr y))
                    (equal (if (ctx? (sub x (car y)))
                               't
                               (ctx? (sub x (cdr y))))
                           't)
                    't)
                't)))
    't)
```

28 Time for If Lifting followed by if-same.

```
(if (ctx? x)
    (if (atom y)
        't
        (if (ctx? (car y))
            't
            (if (ctx? (cdr y))
                (if (equal (ctx? (sub x (cdr y))) 't)
                    (equal (if (ctx? (sub x (car y)))
                               't
                               (ctx? (sub x (cdr y))))
                           't)
                    't)
                't)))
    't)
```

We can use the other inductive premise.

```
(if (ctx? x)
   (if (atom y)
      't
      (if (ctx? (car y))
         't
         (if (ctx? (cdr y))
            (if (equal (ctx? (sub x (cdr y))) 't)
               (equal (if (ctx? (sub x (car y)))
                           't
                           (ctx? (sub x (cdr y))))
                      't)
               't)
            't)))
   't)
```

29 And the rest is quite easily done.

```
't
```

Not quite!

30 That's right,
   we must still prove ctx?/t from
   frame 20.

Here is the claim of ctx?/t. Does this claim need induction?

```
(if (ctx? x)
   (equal (ctx? x) 't)
   't)
```

31 Yes, because ctx? uses recursion over stars.

```
(if (atom x)
   (if (ctx? x)
      (equal (ctx? x) 't)
      't)
   (if (if (ctx? (car x))
          (equal (ctx? (car x)) 't)
          't)
      (if (if (ctx? (cdr x))
             (equal (ctx? (cdr x)) 't)
             't)
         (if (ctx? x)
            (equal (ctx? x) 't)
            't)
         't)
      't))
```

Use ctx? on these two focuses.

```
(if (atom x)
  (if (ctx? x)
    (equal (ctx? x) 't)
    't)
  (if (if (ctx? (car x))
        (equal (ctx? (car x)) 't)
        't)
    (if (if (ctx? (cdr x))
          (equal (ctx? (cdr x)) 't)
          't)
      (if (ctx? x)
        (equal (ctx? x) 't)
        't)
      't)
    't))
```

32 We then use the premise (atom x) and if-nest-A to drop the two resulting ifs.

```
(if (atom x)
  (if (equal x '?)
    (equal (equal x '?) 't)
    't)
  (if (if (ctx? (car x))
        (equal (ctx? (car x)) 't)
        't)
    (if (if (ctx? (cdr x))
          (equal (ctx? (cdr x)) 't)
          't)
      (if (ctx? x)
        (equal (ctx? x) 't)
        't)
      't)
    't))
```

Is this step obvious?

```
(if (atom x)
  (if (equal x '?)
    (equal (equal x† '?) 't)
    't)
  (if (if (ctx? (car x))
        (equal (ctx? (car x)) 't)
        't)
    (if (if (ctx? (cdr x))
          (equal (ctx? (cdr x)) 't)
          't)
      (if (ctx? x)
        (equal (ctx? x) 't)
        't)
      't)
    't))
```

†Hello, there! This is the focus.

33 Yes, indeed.

```
(if (atom x)
  (if (equal x '?)
    (equal (equal '? '?) 't)
    't)
  (if (if (ctx? (car x))
        (equal (ctx? (car x)) 't)
        't)
    (if (if (ctx? (cdr x))
          (equal (ctx? (cdr x)) 't)
          't)
      (if (ctx? x)
        (equal (ctx? x) 't)
        't)
      't)
    't))
```

And this step?

```
(if (atom x)
  (if (equal x '?)
    (equal (equal '? '?) 't)
    't)
  (if (if (ctx? (car x))
        (equal (ctx? (car x)) 't)
        't)
    (if (if (ctx? (cdr x))
          (equal (ctx? (cdr x)) 't)
          't)
      (if (ctx? x)
        (equal (ctx? x) 't)
        't)
      't)
    't))
```

34 This is actually three steps: equal-same, equal-same, and if-same.

```
(if (atom x)
  't
  (if (if (ctx? (car x))
        (equal (ctx? (car x)) 't)
        't)
    (if (if (ctx? (cdr x))
          (equal (ctx? (cdr x)) 't)
          't)
      (if (ctx? x)
        (equal (ctx? x) 't)
        't)
      't)
    't))
```

Now use ctx?, if-nest-E, and the premise written in orange in both focuses.

```
(if (atom x)
  't
  (if (if (ctx? (car x))
        (equal (ctx? (car x)) 't)
        't)
    (if (if (ctx? (cdr x))
          (equal (ctx? (cdr x)) 't)
          't)
      (if (ctx? x)
        (equal (ctx? x) 't)
        't)
      't)
    't))
```

35 Very well.

```
(if (atom x)
  't
  (if (if (ctx? (car x))
        (equal (ctx? (car x)) 't)
        't)
    (if (if (ctx? (cdr x))
          (equal (ctx? (cdr x)) 't)
          't)
      (if (if (ctx? (car x))
            't
            (ctx? (cdr x)))
        (equal (if (ctx? (car x))
                 't
                 (ctx? (cdr x)))
               't)
        't)
      't)
    't))
```

36 What do these three ifs have in common?

```
(if (atom x)
  't
  (if (if (ctx? (car x))
        (equal (ctx? (car x)) 't)
        't)
    (if (if (ctx? (cdr x))
          (equal (ctx? (cdr x)) 't)
          't)
      (if (if (ctx? (car x))
            't
            (ctx? (cdr x)))
        (equal (if (ctx? (car x))
                 't
                 (ctx? (cdr x)))
               't)
        't)
      't)
    't))
```

All three ifs have the question (ctx? (car x)), so we use If Lifting. We also use if-true once where the premise in orange is false, and we use equal-same and if-true at the second orange expression.

```
(if (atom x)
  't
  (if (ctx? (car x))
    (if (equal (ctx? (car x)) 't)
      (if (if (ctx? (cdr x))
            (equal (ctx? (cdr x)) 't)
            't)
        't
        't)
      't)
    (if (if (ctx? (cdr x))
          (equal (ctx? (cdr x)) 't)
          't)
      (if (ctx? (cdr x))
          (equal (ctx? (cdr x)) 't)
          't)
      't)))
```

37 Is this step obvious?

```
(if (atom x)
  't
  (if (ctx? (car x))
    (if (equal (ctx? (car x)) 't)
      (if (if (ctx? (cdr x))
            (equal (ctx? (cdr x)) 't)
            't)
        't
        't)
      't)
    (if (if (ctx? (cdr x))
          (equal (ctx? (cdr x)) 't)
          't)
      (if (ctx? (cdr x))
        (equal (ctx? (cdr x)) 't)
        't)
      't)))
```

Yes, if we remember to use if-same twice.

```
(if (atom x)
  't
  (if (ctx? (car x))
    't
    (if (if (ctx? (cdr x))
          (equal (ctx? (cdr x)) 't)
          't)
      (if (ctx? (cdr x))
        (equal (ctx? (cdr x)) 't)
        't)
      't)))
```

What do these two ifs have in common?

```
(if (atom x)
  't
  (if (ctx? (car x))
    't
    (if (if (ctx? (cdr x))
          (equal (ctx? (cdr x)) 't)
          't)
      (if (ctx? (cdr x))
          (equal (ctx? (cdr x)) 't)
          't)
      't)))
```

38 They have the question (ctx? (cdr x)) in common, so we use If Lifting. We also simplify the final if else using if-same.

```
(if (atom x)
  't
  (if (ctx? (car x))
    't
    (if (ctx? (cdr x))
      (if (equal (ctx? (cdr x)) 't)
          (equal (ctx? (cdr x)) 't)
          't)
      't)))
```

And now?

```
(if (atom x)
  't
  (if (ctx? (car x))
    't
    (if (ctx? (cdr x))
      (if (equal (ctx? (cdr x)) 't)
          (equal (ctx? (cdr x)) 't)
          't)
      't)))
```

39 We use the inductive premise to rewrite (ctx? (cdr x)), and use equal-same to rewrite this focus to 't.

```
(if (atom x)
  't
  (if (ctx? (car x))
    't
    (if (ctx? (cdr x))
      (if (equal (ctx? (cdr x)) 't)
          't
          't)
      't)))
```

What next?

40 We are done proving ctx?/t and ctx?/sub.

Maybe try out Star Induction with J-Bob. The full proofs are on page 188.

41 We might try it out.

8. Learning the Rules

Familiar with this function?

```
(defun member? (x ys)
  (if (atom ys)
      'nil
      (if (equal x (car ys))
          't
          (member? x (cdr ys)))))

measure: (size ys)
```

Define set? that takes a list of atoms and has the value 't if there are no duplicates in the list and otherwise has the value 'nil.

1 It is straightforward with member?.

```
(defun set? (xs)
  (if (atom xs)
      't
      (if (member? (car xs) (cdr xs))
          'nil
          (set? (cdr xs)))))

measure: (size xs)
```

Here is the claim that we need to prove to show that member? is total.

```
(if (natp (size ys))
    (if (atom ys)
        't
        (if (equal x (car ys))
            't
            (< (size (cdr ys)) (size ys))))
    'nil)
```

2 As usual, we drop off the outer if using natp/size and if-true.

```
(if (atom ys)
    't
    (if (equal x (car ys))
        't
        (< (size (cdr ys)) (size ys))))
```

Then we use size/cdr and the premise (atom ys).

```
(if (atom ys)
    't
    (if (equal x (car ys))
        't
        (< (size (cdr ys)) (size ys))))
```

3 That's familiar, too.

```
(if (atom ys)
    't
    (if (equal x (car ys))
        't
        't))
```

And then we are done.

```
(if (atom ys)
    't
    (if (equal x (car ys))
        't
        't))
```

4 By if-same twice.

```
't
```

---

5 Now for the claim that set? is total. The proof uses the identical sequence of six steps as that of member?. How do the totality proofs of set? and member? differ?

```
(if (natp (size xs))
   (if (atom xs)
      't
      (if (member? (car xs) (cdr xs))
         't
         (< (size (cdr xs)) (size xs))))
   'nil)
```

The expression (equal x (car ys)) is replaced by (member? (car xs) (cdr xs)), but neither proof cares about that if question.

Q.E.D.

't

---

6 Here is the definition of atoms. Can we state and prove a claim about it?

```
(defun atoms (x)
  (add-atoms x '()))
```

No, because we do not know what add-atoms is.

---

7 Here is the definition of add-atoms. Can we prove a claim about atoms now?

```
(defun add-atoms (x ys)
  (if (atom x)
      (if (member? x ys)
          ys
          (cons x ys))
      (add-atoms (car x)
        (add-atoms (cdr x) ys))))

measure: (size x)
```

No, we don't yet know whether add-atoms is total.

---

8 What do we need to know to determine whether add-atoms is total?

We need to know its totality claim.

---

9 Good point. Does add-atoms have a totality claim?

We do not know.

What does a totality claim tell us about a function?

10 A totality claim tells us that the function's measure decreases on every recursive call.

Exactly. Can we state add-atoms's claim?

11 Perhaps.

What is add-atoms's measure?

12 Its measure is (size x).

Does add-atoms have a recursive application?

13 Yes,

```
(add-atoms (car x)
  (add-atoms (cdr x) ys)).
```

We must state that the measure (size x) decreases for

```
(add-atoms (car x)
  (add-atoms (cdr x) ys)).
```

What does it mean for the measure to decrease for this recursive application?

14 We are not sure.

How would we use the Law of Defun for the recursive application

```
(add-atoms (car x)
  (add-atoms (cdr x) ys))?
```

15 By replacing x with (car x) and ys with (add-atoms (cdr x) ys) in the body of add-atoms.

Let's do the same for the measure of add-atoms.

16 As in, replace x with (car x) and ys with (add-atoms (cdr x) ys) in (size x)?

Correct.

17 The result is (size (car x)).

Exactly. The measure for the recursive application

```
(add-atoms (car x)
  (add-atoms (cdr x) ys))
```

is (size (cdr x)). What does it mean for the measure to decrease for this recursive application?

18 Perhaps it means that the measure for this recursive application is less than the measure for add-atoms.

How do we state that claim as an expression?

19 Here is the expression: (< (size (car x)) (size x)). Is this add-atoms's totality claim?

No, not yet. We still have one recursive application left.

20 Yes, the second argument of the first recursive application in add-atoms is also recursive: (add-atoms (cdr x) ys).

What is the measure for the remaining recursive application from add-atoms?

21 By replacing x with (cdr x) and ys with ys in (size x), we get (size (cdr x)).

What next?

22 Next, we state the claim that the measure decreases for the remaining recursive application: (< (size (cdr x)) (size x)).

Yes. We have separate statements for each recursive application in add-atoms that the measure must decrease. State that both claims must be true.

23 That's easy.

```
(if (< (size (car x)) (size x))
   (< (size (cdr x)) (size x))
   'nil)
```

Is this add-atoms's totality claim?

No, that is still not the complete totality claim. Are we sure the expression in frame 23 states that both (< (size (car x)) (size x)) and (< (size (cdr x)) (size x)) must be true?

24 Of course.

Conjunction

The *conjunction* of expressions $e_1 \ldots e_n$ states that each of $e_1, \ldots, e_n$ must be true.

The conjunction of zero expressions is 't.

The conjunction of one expression e_1 is e_1.

The conjunction of e_1 and e_2 is e_1 if e_2 is 't, it is e_2 if e_1 is 't, and otherwise it is (if e_1 e_2 'nil).

The conjunction of three or more expressions e_1 e_2 $\ldots$ e_n is the conjunction of e_1 and the conjunction of e_2 $\ldots$ e_n.

What does the claim in frame 23 state?

25 It states that the measure of add-atoms must decrease for the function's two recursive applications.

When does add-atoms call itself recursively?

26 When x is not an atom.

That's right. For add-atoms to be total, we only need to prove the claim in frame 23 in the case where x is not an atom. Restate the claim to say so.

27 What about the case where x is an atom?

There are no recursive applications in add-atoms when x is an atom, so there's nothing we need to prove.

28 In that case, here is the new claim.

```
(if (atom x)
  't
  (if (< (size (car x)) (size x))
    (< (size (cdr x)) (size x))
    'nil))
```

This is the totality claim for add-atoms, isn't it?

Unfortunately not. But we are almost finished.

29 Good. What's next?

We have stated that the measure for add-atoms must decrease when it makes recursive calls. How many times can it decrease?

30 At most (size x), surely.

What prevents the measure from decreasing more than (size x) times?

31 Ah, of course!

```
(if (natp (size x))
  (if (atom x)
    't
    (if (< (size (car x)) (size x))
      (< (size (cdr x)) (size x))
      'nil))
  'nil)
```

Constructing Totality Claims

Given a function (defun *name* (x_1 ... x_n) *body*) and a measure m, construct a claim for subexpressions in *body*:

For variables and quoted literals, use 't.

For (if Q A E) where the claims for Q, A, and E are c_q, c_a, and c_e, if c_a and c_e are the same use the conjunction of c_q and c_a, otherwise use the conjunction of c_q and (if Q c_a c_e).

For any other expression E, consider each recursive application (*name* e_1 ... e_n) in E. Construct the measure m_r of the recursive application by substituting e_1 for x_1, ..., e_n for x_n in m. The claim for E is the conjunction of ($<$ m_r m) for every recursive application in E.

The totality claim for *name* is the conjunction of (natp m) and the claim for *body*.

And *that* is the totality claim for add-atoms.

32 Finally.

We can prove the totality claim for add-atoms.

```
(if (natp (size x))
  (if (atom x)
    't
    (if (< (size (car x)) (size x))
      (< (size (cdr x)) (size x))
      'nil))
  'nil)
```

33 Easily done.

```
't
```

Can we prove something about add-atoms now?

First let's take a break. We can walk through this totality proof with J-Bob on page 192, and find out how J-Bob constructs totality claims in Appendix C.

34 Or perhaps it is just time for a mid-morning snack.

9. Changing the Rules

Now that we know add-atoms and atoms are total, can we state and prove a claim about either of them?	1	No, we do not understand the definition of add-atoms.
Given a value as its first argument and a list of unique atoms as its second argument, add-atoms adds all of the atoms from the first argument to the second argument, except those that are already in the list. What value is (add-atoms '(a . (b . (c . a))) '(d a e))[†] equal to? [†]The notation (x . y) inside a quoted literal represents the result of (cons 'x 'y). Normally, we only use cons when its second argument is a list. In this chapter and in chapter 10, however, we ignore this restriction in order to write simpler functions and proofs.	2	'(b c d a e).
What value is (add-atoms '((a . b) . (c . a)) '(d a e)) equal to?	3	'(b c d a e).
What value is (add-atoms 'a '(d a e)) equal to?	4	'(d a e).
What value is (add-atoms 'b '(d a e)) equal to?	5	'(b d a e).
What value is (atoms '(((a . b) . (c . a)) . (d . (a . e)))) equal to?	6	'(b c d a e).

We can now state a claim about atoms.

```
(dethm set?/atoms (a)
  (equal (set? (atoms a)) 't))
```

7 Let's try to prove it.

What do we do with this focus?

```
(equal (set? (atoms a)) 't)
```

8 We start by using atoms.

```
(equal (set? (add-atoms a '())) 't)
```

What now?

9 The function add-atoms treats a as either an atom or as nested cons pairs. So, we can use Star Induction.

In order to use induction, let's create a separate theorem about add-atoms.

10 Certainly.

```
(dethm set?/add-atoms (a)
  (equal (set? (add-atoms a '())) 't))
```

What is the inductive claim we must prove?

```
(equal (set? (add-atoms a '())) 't)
```

11 Here it is.

```
(if (atom a)
  (equal (set? (add-atoms a '())) 't)
  (if (equal (set? (add-atoms (car a) '())) 't)
    (if (equal (set? (add-atoms (cdr a) '())) 't)
      (equal (set? (add-atoms a '())) 't)
      't)
    't))
```

Let's use add-atoms on a and '(). Do the recursive applications correspond to our inductive premises?

```
(if (atom a)
  (equal (set? (add-atoms a '())) 't)
  (if (equal (set? (add-atoms (car a) '())) 't)
    (if (equal (set? (add-atoms (cdr a) '())) 't)
      (equal (set?
               (add-atoms a '()))
             't)
      't)
    't))
```

12 No, they do not; the inductive premise for (car a) and the recursive application for (car a) do not match. In the inductive premise, the second argument is (add-atoms (cdr a) '()). In the recursive application, the second argument is '().

```
(if (atom a)
  (equal (set? (add-atoms a '())) 't)
  (if (equal (set? (add-atoms (car a) '())) 't)
    (if (equal (set? (add-atoms (cdr a) '())) 't)
      (equal (set?
               (if (atom a)
                 (if (member? a '())
                   '()
                   (cons a '()))
                 (add-atoms (car a)
                   (add-atoms (cdr a) '()))))
             't)
      't)
    't))
```

Star Induction is not helping us in this proof. We need inductive premises that match the recursion in our functions.

13 We only know List Induction and Star Induction, and neither of them look like the recursion in add-atoms.

Does add-atoms use natural recursion?

14 Yes, for only one argument. The argument x is replaced by (car x) and (cdr x) in the recursive applications. The argument ys, however, stays ys in one recursive application and becomes (add-atoms (cdr x) ys) in the other.

Why isn't

```
(add-atoms (car x)
  (add-atoms (cdr x) ys))
```

a natural recursion?

15 In a natural recursion, every argument must stay the same or be a structural part of the given one. The argument (add-atoms (cdr x) ys) is neither the same as ys, nor a part of ys. If add-atoms does not use natural recursion, can we prove set?/atoms by induction?

Induction uses inductive premises, which are natural recursions for proofs. Without natural recursion, our proof does not have the right inductive premises. We are stuck.

16 Is there hope?

Yes, but we need to find a kind of induction that matches the recursion in add-atoms.

17 Is there a kind of induction like that?

We can *create* a new kind of induction. Given any total, recursive function, we can always state inductive claims based on the recursion in the function.

18 That's surprising. Does that mean we can create a whole new kind of induction just for add-atoms?

Absolutely. To begin, we need to state a claim in terms of add-atoms that we want to use our new kind of induction on.

19 How about the claim in frame 11?

```
(equal (set? (add-atoms a '())) 't)
```

That's a start, but in order for induction to substitute the correct inductive premises, we need the arguments of add-atoms to be variables. Can we state a more general claim where the arguments are both variables?

20 We certainly can.

```
(equal (set? (add-atoms a bs)) 't)
```

Is that claim true?

21 No, (equal (set? (add-atoms 'a '(b b))) 't) is a counterexample.

What must be true of bs for the claim to hold?

22 If bs is a list with no duplicates, then (add-atoms a bs) is also a list with no duplicates.

Can we state that claim?

23 Certainly.

```
(dethm set?/add-atoms (a bs)
  (if (set? bs)
      (equal (set? (add-atoms a bs)) 't)
      't))
```

Now how does induction for add-atoms work?

What must the inductive claim for set?/add-atoms state?

24 Good question. We do not know.

The inductive claim for set?/add-atoms must state that in the cases where add-atoms does not call itself, set?/add-atoms is true, and in the cases where add-atoms does call itself, set?/add-atoms is true of the original arguments if it is also true of the arguments to the recursive applications.

25 How do we state that?

In what case is add-atoms non-recursive?

26 When x is an atom, add-atoms is non-recursive.

In that case, we must state that set?/add-atoms is true.

27 In other words, we must state

```
(if (set? bs)
    (equal (set? (add-atoms a bs)) 't)
    't).
```

That's easy so far.

Precisely. What case remains?

28 The case where x is not an atom.

In that case, we must also state that set?/add-atoms is true.

29 Can inductive claims really be this simple?

```
(if (set? bs)
  (equal (set? (add-atoms a bs)) 't)
  't)
```

Not quite. We must still address recursive applications in add-atoms. What is the first recursive application?

30

```
(add-atoms (car x)
  (add-atoms (cdr x) ys)).
```

We must state that set?/add-atoms holds for the arguments of this recursive application.

31 In other words, replace x with (car x) and ys with (add-atoms (cdr x) ys) in set?/add-atoms.

Almost. In set?/add-atoms, do x and ys appear?

32 No, they do not.

An inductive claim does not necessarily use the same variables as the recursive function it is based on. Recall that in frame 20 we do choose to use variables as the arguments to add-atoms in set?/add-atoms. What variables do we use?

33 We use a and bs.

What is the first recursive application in add-atoms, if we use a and bs instead of x and ys?

34

```
(add-atoms (car a)
  (add-atoms (cdr a) bs)).
```

How do we state that set?/add-atoms holds for the arguments of *this* recursive application?

35 We replace a with (car a) and bs with (add-atoms (cdr a) bs) in set?/add-atoms.

Very good. What is the result?

36
```
(if (set? (add-atoms (cdr a) bs))
    (equal (set? (add-atoms (car a)
                            (add-atoms (cdr a) bs)))
           't)
    't)
```

What next?

The recursive application (add-atoms (cdr a) bs) remains.

37 Presumably, we must replace a with (car a) and bs with bs in set?/add-atoms.

That presumption is correct, indeed.

38
```
(if (set? bs)
    (equal (set? (add-atoms (cdr a) bs))
           't)
    't)
```

What is this expression for?

The expressions in frames 36 and 38 are the inductive premises for the inductive claim of set?/add-atoms.

39 Interesting.

Inductive Premises

Given a claim c, a recursive application ($name\ e_1 \ldots e_n$), and variables $x_1 \ldots x_n$, the inductive premise for this application is c where x_1 is e_1, ..., x_n is e_n.

Next, state that if both inductive premises are true, set?/add-atoms must be true.

40

```
(if (if (set? (add-atoms (cdr a) bs))
        (equal (set? (add-atoms (car a)
                                (add-atoms (cdr a) bs)))
               't)
        't)
    (if (if (set? bs)
            (equal (set? (add-atoms (cdr a) bs))
                   't)
            't)
        (if (set? bs)
            (equal (set? (add-atoms a bs))
                   't)
            't)
        't)
    't).
```

Are we sure that this expression states that if the inductive premises in frames 36 and 38 are true, then set?/add-atoms must be true as well?

41 Indubitably.

Implication

An *implication* states that some premises *imply* a conclusion. In other words, when the premises $e_1 \ldots e_n$ are true, the conclusion e_0 must be true as well.

For zero premises, the implication is e_0.

For one premise e_1, the implication is (if e_1 e_0 't).

For two or more premises e_1 e_2 $\ldots$ e_n, state that e_1 implies that the conjunction of the premises $e_2 \ldots e_n$ imply the conclusion e_0.

In frames 27 and 40, we state claims about set?/add-atoms for the cases where a is an atom and where a is not an atom.

42 What is left to do?

Put the two cases together, of course.

43 We can certainly do that:

```
(if (atom a)
  (if (set? bs)
    (equal (set? (add-atoms a bs)) 't)
    't)
  (if (if (set? (add-atoms (cdr a) bs))
        (equal (set? (add-atoms (car a)
                                (add-atoms (cdr a) bs)))
               't)
        't)
    (if (if (set? bs)
          (equal (set? (add-atoms (cdr a) bs))
                 't)
          't)
        (if (set? bs)
          (equal (set? (add-atoms a bs)) 't)
          't)
        't)
    't)).
```

Defun Induction

Given a claim c, a function (defun $name_f$ $(x_1 \ldots x_n)$ $body_f$), and a choice of variables $y_1 \ldots y_n$, we construct a claim for subexpressions of $body_i$, where $body_i$ is $body_f$ with x_1 replaced by y_1, ..., and x_n replaced by y_n:

For (if Q A E) where the claims for A and E are c_a and c_e, state that the inductive premises of Q imply c_{ae}, where c_{ae} is c_a if c_a is equal to c_e, and (if Q c_a c_e) otherwise.

For any other expression E, state that the inductive premises of E imply c.

The inductive claim of c is the claim for $body_i$.

That is the inductive claim for set?/add-atoms, using induction based on the definition of add-atoms.

44 There is nothing to it!

Really?

45 No, not really.

We have stated the claim to prove set?/add-atoms by Defun Induction using add-atoms.

46 Impressive! How does this differ from List Induction or Star Induction?

Actually, these forms of induction do not differ. List Induction and Star Induction are each based on Defun Induction. We define a function for each that produces the appropriate inductive premises.

47 What functions are those?

We use list-induction for List Induction.

48 How about Star Induction?

```
(defun list-induction (x)
  (if (atom x)
      '()
      (cons (car x)
        (list-induction (cdr x)))))

measure: (size x)
```

We use star-induction, of course.

49 Are there other function definitions we should use for common forms of induction?

```
(defun star-induction (x)
  (if (atom x)
      x
      (cons (star-induction (car x))
        (star-induction (cdr x)))))

measure: (size x)
```

None come to mind, but they have a way of popping up when we least expect them.

50 Does Defun Induction work for every function we write?

The ways that we construct inductive claims and totality claims work for every function, but they work better for some functions than others.

51 Which functions are those?

When we construct inductive claims and totality claims, we assume that if expressions are outside all other expressions; we ignore if expressions in the arguments of applications.

52 What happens when an if expression is inside the argument of an application?

In some cases we might get a claim we cannot prove, even though our defun is total or our dethm is actually a theorem.

53 What can we do then?

There are trickier ways to create totality claims and inductive claims that work for all ifs. But we can always just rewrite our functions to have the ifs on the outside.

54 How?

With If Lifting, of course!

55 Of course. *Then* can we write totality claims and inductive claims that work for every total function?

No, not every total function.[†] There is always more to learn.

[†] Some total functions do not have a natural number measure. Many more functions can be proved total by using *ordinal numbers*. See "Restless for More?" for further reading.

56 Now back to the proof, please!

We should try to prove set?/add-atoms again using Defun Induction on add-atoms.

```
(dethm set?/add-atoms (a bs)
  (if (set? bs)
      (equal (set? (add-atoms a bs)) 't)
      't))
```

57 And here again is the inductive claim from frame 43.

```
(if (atom a)
    (if (set? bs)
        (equal (set? (add-atoms a bs)) 't)
        't)
    (if (if (set? (add-atoms (cdr a) bs))
            (equal (set? (add-atoms (car a)
                                    (add-atoms (cdr a) bs)))
                   't)
            't)
        (if (if (set? bs)
                (equal (set? (add-atoms (cdr a) bs)) 't)
                't)
            (if (set? bs)
                (equal (set? (add-atoms a bs)) 't)
                't)
            't)
        't))
```

Next, we use add-atoms, if-nest-A, and the premise (atom a).

```
(if (atom a)
    (if (set? bs)
        (equal (set? (add-atoms a bs))
               't)
        't)
    (if (if (set? (add-atoms (cdr a) bs))
            (equal (set? (add-atoms (car a)
                                    (add-atoms (cdr a) bs)))
                   't)
            't)
        (if (if (set? bs)
                (equal (set? (add-atoms (cdr a) bs)) 't)
                't)
            (if (set? bs)
                (equal (set? (add-atoms a bs)) 't)
                't)
            't)
        't))
```

58 Yes, these steps are easy.

```
(if (atom a)
    (if (set? bs)
        (equal (set? (if (member? a bs)
                         bs
                         (cons a bs)))
               't)
        't)
    (if (if (set? (add-atoms (cdr a) bs))
            (equal (set? (add-atoms (car a)
                                    (add-atoms (cdr a) bs)))
                   't)
            't)
        (if (if (set? bs)
                (equal (set? (add-atoms (cdr a) bs)) 't)
                't)
            (if (set? bs)
                (equal (set? (add-atoms a bs)) 't)
                't)
            't)
        't))
```

59 What comes next?

```
(if (atom a)
   (if (set? bs)
      (equal (set? (if (member? a bs)
                       bs
                       (cons a bs)))
             't)
      't)
   (if (if (set? (add-atoms (cdr a) bs))
          (equal (set? (add-atoms (car a)
                          (add-atoms (cdr a) bs)))
                 't)
          't)
      (if (if (set? bs)
             (equal (set? (add-atoms (cdr a) bs)) 't)
             't)
          (if (set? bs)
             (equal (set? (add-atoms a bs)) 't)
             't)
          't)
      't))
```

If Lifting on (member? a bs).

```
(if (atom a)
   (if (set? bs)
      (equal (if (member? a bs)
                 (set? bs)
                 (set? (cons a bs)))
             't)
      't)
   (if (if (set? (add-atoms (cdr a) bs))
          (equal (set? (add-atoms (car a)
                          (add-atoms (cdr a) bs)))
                 't)
          't)
      (if (if (set? bs)
             (equal (set? (add-atoms (cdr a) bs)) 't)
             't)
          (if (set? bs)
             (equal (set? (add-atoms a bs)) 't)
             't)
          't)
      't))
```

60 If (set? bs) is true in the premise, then it should be equal to 't, right?

```
(if (atom a)
   (if (set? bs)
      (equal (if (member? a bs)
                 (set? bs)
                 (set? (cons a bs)))
             't)
      't)
   (if (if (set? (add-atoms (cdr a) bs))
          (equal (set? (add-atoms (car a)
                          (add-atoms (cdr a) bs)))
                 't)
          't)
      (if (if (set? bs)
             (equal (set? (add-atoms (cdr a) bs)) 't)
             't)
          (if (set? bs)
             (equal (set? (add-atoms a bs)) 't)
             't)
          't)
      't))
```

That seems reasonable.

```
(if (atom a)
   (if (set? bs)
      (equal (if (member? a bs)
                 't
                 (set? (cons a bs)))
             't)
      't)
   (if (if (set? (add-atoms (cdr a) bs))
          (equal (set? (add-atoms (car a)
                          (add-atoms (cdr a) bs)))
                 't)
          't)
      (if (if (set? bs)
             (equal (set? (add-atoms (cdr a) bs)) 't)
             't)
          (if (set? bs)
             (equal (set? (add-atoms a bs)) 't)
             't)
          't)
      't))
```

How can we do that?

61 With the claim set?/t.

```
(dethm set?/t (xs)
  (if (set? xs)
      (equal (set? xs) 't)
      't))
```

Like ctx?/t in frame 19 of chapter 7, we state set?/t as a separate claim, and we must eventually prove the claim.

Now we use set? with the argument (cons a bs).

```
(if (atom a)
    (if (set? bs)
        (equal (if (member? a bs)
                   't
                   (set? (cons a bs)))
               't)
        't)
    (if (if (set? (add-atoms (cdr a) bs))
            (equal (set? (add-atoms (car a)
                                    (add-atoms (cdr a) bs)))
                   't)
            't)
        (if (if (set? bs)
                (equal (set? (add-atoms (cdr a) bs)) 't)
                't)
            (if (set? bs)
                (equal (set? (add-atoms a bs)) 't)
                't)
            't)
        't))
```

62 We simplify the result using atom/cons, car/cons, cdr/cons twice, and if-false.

```
(if (atom a)
    (if (set? bs)
        (equal (if (member? a bs)
                   't
                   (if (member? a bs)
                       'nil
                       (set? bs)))
               't)
        't)
    (if (if (set? (add-atoms (cdr a) bs))
            (equal (set? (add-atoms (car a)
                                    (add-atoms (cdr a) bs)))
                   't)
            't)
        (if (if (set? bs)
                (equal (set? (add-atoms (cdr a) bs)) 't)
                't)
            (if (set? bs)
                (equal (set? (add-atoms a bs)) 't)
                't)
            't)
        't))
```

It looks like our premises (set? bs) and (member? a bs) are about to come in handy.

Then let's use those premises.

```
(if (atom a)
  (if (set? bs)
    (equal (if (member? a bs)
                'ᵗ
                (if (member? a bs)
                  'nil
                  (set? bs)))
          't)
    't)
  (if (if (set? (add-atoms (cdr a) bs))
        (equal (set? (add-atoms (car a)
                                (add-atoms (cdr a) bs)))
              't)
        't)
    (if (if (set? bs)
          (equal (set? (add-atoms (cdr a) bs)) 't)
          't)
      (if (set? bs)
        (equal (set? (add-atoms a bs)) 't)
        't)
      't)
    't))
```

63 And with a few more rewrites, we finish the case where a is an atom.

```
(if (atom a)
  't
  (if (if (set? (add-atoms (cdr a) bs))
        (equal (set? (add-atoms (car a)
                                (add-atoms (cdr a) bs)))
              't)
        't)
    (if (if (set? bs)
          (equal (set? (add-atoms (cdr a) bs)) 't)
          't)
      (if (set? bs)
        (equal (set? (add-atoms a bs)) 't)
        't)
      't)
    't))
```

Do these two ifs have anything in common?

```
(if (atom a)
  't
  (if (if (set? (add-atoms (cdr a) bs))
        (equal (set? (add-atoms (car a)
                                (add-atoms (cdr a) bs)))
              't)
        't)
    (if (if (set? bs)
          (equal (set? (add-atoms (cdr a) bs)) 't)
          't)
      (if (set? bs)
        (equal (set? (add-atoms a bs)) 't)
        't)
      't)
    't))
```

64 Certainly, they have the same if question. We can combine them using If Lifting, of course. We use if-same twice in the final if else, as well.

```
(if (atom a)
  't
  (if (set? bs)
    (if (if (set? (add-atoms (cdr a) bs))
          (equal (set? (add-atoms (car a)
                                  (add-atoms (cdr a) bs)))
                't)
          't)
      (if (equal (set? (add-atoms (cdr a) bs)) 't)
        (equal (set? (add-atoms a bs)) 't)
        't)
      't)
    't))
```

We use If Lifting again, since both if questions essentially ask (set? (add-atoms (cdr a) bs)).

```
(if (atom a)
  't
  (if (set? bs)
    (if (if (set? (add-atoms (cdr a) bs))
          (equal (set? (add-atoms (car a)
                                  (add-atoms (cdr a) bs)))
                 't)
          't)
        (if (equal (set? (add-atoms (cdr a) bs)) 't)
          (equal (set? (add-atoms a bs)) 't)
          't)
        't)
    't))
```

65 We also simplify the else of the new if using if-true.

```
(if (atom a)
  't
  (if (set? bs)
    (if (set? (add-atoms (cdr a) bs))
      (if (equal (set? (add-atoms (car a)
                                  (add-atoms (cdr a) bs)))
                 't)
        (if (equal (set? (add-atoms (cdr a) bs)) 't)
          (equal (set? (add-atoms a bs)) 't)
          't)
        't)
      (if (equal (set? (add-atoms (cdr a) bs)) 't)
        (equal (set? (add-atoms a bs)) 't)
        't))
    't))
```

The if questions written in orange both mean the same thing as (set? (add-atoms (cdr a) bs)), but we can't simplify them using if-nest-A or if-nest-E. Do we have a way to rewrite these?

```
(if (atom a)
  't
  (if (set? bs)
    (if (set? (add-atoms (cdr a) bs))
      (if (equal (set? (add-atoms (car a)
                                  (add-atoms (cdr a) bs)))
                 't)
        (if (equal (set? (add-atoms (cdr a) bs)) 't)
          (equal (set? (add-atoms a bs)) 't)
          't)
        't)
      (if (equal (set? (add-atoms (cdr a) bs)) 't)
        (equal (set? (add-atoms a bs)) 't)
        't))
    't))
```

66 In the case where the premise (set? (add-atoms (cdr a) bs)) is true, it is equal to 't by set?/t. In the case where the premise is false, it is equal to 'nil.

In that case, we need a new claim to say that it is equal to 'nil.

67 Naturally.

```
(dethm set?/nil (xs)
  (if (set? xs)
      't
      (equal (set? xs) 'nil)))
```

Let's use our new theorem.

```
(if (atom a)
  't
  (if (set? bs)
    (if (set? (add-atoms (cdr a) bs))
      (if (equal (set? (add-atoms (car a)
                                  (add-atoms (cdr a) bs)))
                 't)
        (if (equal (set? (add-atoms (cdr a) bs)) 't)
          (equal (set? (add-atoms a bs)) 't)
          't)
        't)
      (if (equal (set? (add-atoms (cdr a) bs)) 't)
        (equal (set? (add-atoms a bs)) 't)
        't))
    't))
```

68 Very well. We shall use set?/t and set?/nil with the premise written in orange.

```
(if (atom a)
  't
  (if (set? bs)
    (if (set? (add-atoms (cdr a) bs))
      (if (equal (set? (add-atoms (car a)
                                  (add-atoms (cdr a) bs)))
                 't)
        (if (equal 't 't)
          (equal (set? (add-atoms a bs)) 't)
          't)
        't)
      (if (equal 'nil 't)
        (equal (set? (add-atoms a bs)) 't)
        't))
    't))
```

Simplify both ifs.

```
(if (atom a)
  't
  (if (set? bs)
    (if (set? (add-atoms (cdr a) bs))
      (if (equal (set? (add-atoms (car a)
                                  (add-atoms (cdr a) bs)))
                 't)
        (if (equal 't 't)
          (equal (set? (add-atoms a bs)) 't)
          't)
        't)
      (if (equal 'nil 't)
        (equal (set? (add-atoms a bs)) 't)
        't))
    't))
```

69 Easy, using **equal** twice, **if-true**, and **if-false**.

```
(if (atom a)
  't
  (if (set? bs)
    (if (set? (add-atoms (cdr a) bs))
      (if (equal (set? (add-atoms (car a)
                                  (add-atoms (cdr a) bs)))
                 't)
        (equal (set? (add-atoms a bs)) 't)
        't)
      't)
    't))
```

Use **add-atoms**.

```
(if (atom a)
  't
  (if (set? bs)
    (if (set? (add-atoms (cdr a) bs))
      (if (equal (set? (add-atoms (car a)
                                  (add-atoms (cdr a) bs)))
                 't)
        (equal (set? (add-atoms a bs))
               't)
        't)
      't)
    't))
```

70 We also drop an **if** using **if-nest-E** and the outermost premise.

```
(if (atom a)
  't
  (if (set? bs)
    (if (set? (add-atoms (cdr a) bs))
      (if (equal (set? (add-atoms (car a)
                                  (add-atoms (cdr a) bs)))
                 't)
        (equal (set? (add-atoms (car a)
                                (add-atoms (cdr a) bs)))
               't)
        't)
      't)
    't))
```

And now?

```
(if (atom a)
  't
  (if (set? bs)
    (if (set? (add-atoms (cdr a) bs))
      (if (equal (set? (add-atoms (car a)
                        (add-atoms (cdr a) bs)))
             't)
        (equal (set? (add-atoms (car a)
                        (add-atoms (cdr a) bs)))
             't)
        't)
      't)
    't))
```

71 We can finally use the first inductive premise.

```
(if (atom a)
  't
  (if (set? bs)
    (if (set? (add-atoms (cdr a) bs))
      (if (equal (set? (add-atoms (car a)
                        (add-atoms (cdr a) bs)))
             't)
        't
        't)
      't)
    't))
```

And then we are essentially done.

Let's return to set?/atoms. Are we ready to prove it?

```
(dethm set?/atoms (a)
  (equal (set? (atoms a)) 't))
```

72 Hopefully.

This proof is a little surprising, a little fun, and a little proof.

73 That sounds a little exciting.

First, use atoms.

```
(equal (set? (atoms a)) 't)
```

74 Very well. Can we use set?/add-atoms here?

```
(equal (set? (add-atoms a '())) 't)
```

Not yet, we need a premise. if-true makes room for one.

```
(equal (set? (add-atoms a '())) 't)
```

Surprise number one: if-true yields a premise wherever we need one.

75 We need this premise to be (set? '()). How can we rewrite 't to (set? '())?

```
(if 't
  (equal (set? (add-atoms a '())) 't)
  't)
```

The if-true theorem rewrites 't to an if that has the same answer and else as the body of set?, where xs is '().

```
(if 't
  (equal (set? (add-atoms a '())) 't)
  't)
```

Surprise number two: if-true lets us supply any else expression we like.

76 And the question of the if?

```
(if (if 't
        't
        (if (member? (car '()) (cdr '()))
            'nil
            (set? (cdr '()))))
  (equal (set? (add-atoms a '())) 't)
  't)
```

Using atom rewrites 't to the if question from set?, again where xs is '().

```
(if (if 't
        't
        (if (member? (car '()) (cdr '()))
            'nil
            (set? (cdr '()))))
  (equal (set? (add-atoms a '())) 't)
  't)
```

Surprise number three: we can "run" atom in reverse, since 't and (atom '()) are equal.

77 And now?

```
(if (if (atom '())
        't
        (if (member? (car '()) (cdr '()))
            'nil
            (set? (cdr '()))))
  (equal (set? (add-atoms a '())) 't)
  't)
```

We use set?, where xs is '()!

```
(if (if (atom '())
        't
        (if (member? (car '()) (cdr '()))
            'nil
            (set? (cdr '()))))
  (equal (set? (add-atoms a '())) 't)
  't)
```

The fourth and final surprise: we can use the Law of Defun to rewrite the body of set? where xs is '() to an application of set?.

78 And now we can use set?/add-atoms.

```
(if (set? '())
  (equal (set? (add-atoms a '())) 't)
  't)
```

Exactly.

```
(if (set? '())
    (equal (set? (add-atoms a '())) 't)
    't)
```

79 And now we are virtually done.

```
't
```

Wasn't that a little fun?

80 And quite surprising.

Do we need to prove set?/t and set?/nil?

81 Certainly we must.

Try them out with J-Bob on page 193.

82 Can we try Defun Induction with J-Bob as well?

Of course. Appendix A tells how to use J-Bob for Defun Induction, and Appendix C shows how J-Bob constructs inductive claims.

83 That is good to know.

Now we know everything there is to learn about Defun Induction.

84 Well, perhaps not everything.

Go back through this chapter if necessary. Defun Induction can take time.

85 It certainly can.

10. The Stars Are Aligned

What value is

```
(cons
  (cons
    (cons
      (cons 'french 'toast)
      'and)
    'maple)
  'syrup)
```

equal to?

1 '((((french . toast) . and) . maple) . syrup).

How many conses are in

'((((french . toast) . and) . maple) . syrup)?

2 Four, since each "." counts for one cons.

What value is

```
(rotate
  '((((french . toast) . and) . maple) . syrup))
```

equal to?

3 '(((french . toast) . and) . (maple . syrup)).

How many conses are in

'(((french . toast) . and) . (maple . syrup))?

4 Four again.

What value is

```
(rotate
  '(((french . toast) . and) . (maple . syrup))).
```

equal to?

5 '((french . toast) . (and . (maple . syrup))).

How many conses are in '((french . toast) . (and . (maple . syrup)))?

6 Still four.

What value is
(rotate
  '((french . toast) . (and . (maple . syrup)))).
equal to?

7 '(french . (toast . (and . (maple . syrup)))).

How many conses are in '(french . (toast . (and . (maple . syrup))))?

8 Yet again, four!

Define rotate.

9

```
(defun rotate (x)
  (cons (car (car x))
    (cons (cdr (car x)) (cdr x))))
```

Is rotate/cons a theorem?

```
(dethm rotate/cons (x y z)
  (equal (rotate (cons (cons x y) z))
         (cons x (cons y z))))
```

10 We can find out.

Use rotate.

```
(equal (rotate (cons (cons x y) z))
       (cons x (cons y z)))
```

11 The claim is starting to look complicated.

```
(equal (cons (car (car (cons (cons x y) z)))
             (cons (cdr (car (cons (cons x y) z)))
               (cdr (cons (cons x y) z))))
       (cons x (cons y z)))
```

12 Simplify with car/cons and cdr/cons.

```
(equal (cons (car (car (cons (cons x y) z)))
             (cons (cdr (car (cons (cons x y) z)))
                   (cdr (cons (cons x y) z))))
       (cons x (cons y z)))
```

Much better.

```
(equal (cons x
             (cons y
                   z))
       (cons x (cons y z)))
```

13 And now ...

```
(equal (cons x (cons y z))
       (cons x (cons y z)))
```

We are done!

```
't
```

14 What value is

```
(align
  '(french . (toast . (and . (maple . syrup)))))
```

equal to?

'(french . (toast . (and . (maple . syrup)))).

15 What value is

```
(align
  '((french . toast) . (and . (maple . syrup))))
```

equal to?

'(french . (toast . (and . (maple . syrup)))).

16 What value is

```
(align
  '(((french . toast) . and) . (maple . syrup)))
```

equal to?

'(french . (toast . (and . (maple . syrup)))).

17 What value is

```
(align
  '((((french . toast) . and) . maple) . syrup))
```

equal to?

'(french . (toast . (and . (maple . syrup)))).

Define align.

18

```
(defun align (x)
  (if (atom x)
      x
      (if (atom (car x))
          (cons (car x) (align (cdr x)))
          (align (rotate x)))))

measure: (size x)
```

Is (size x) an appropriate measure for align?

19 The measure (size x) has always worked.

Then let's see if we can prove that align is total. Of course, (natp (size x)) is 't.

```
(if (natp (size x))
  (if (atom x)
      't
      (if (atom (car x))
          (< (size (cdr x)) (size x))
          (< (size (rotate x)) (size x))))
  'nil)
```

20 Yes, and by if-true, we remove the outer if.

```
(if (atom x)
    't
    (if (atom (car x))
        (< (size (cdr x)) (size x))
        (< (size (rotate x)) (size x))))
```

Use size/cdr and the premise (atom x).

```
(if (atom x)
    't
    (if (atom (car x))
        (< (size (cdr x)) (size x))
        (< (size (rotate x)) (size x))))
```

21 Simple.

```
(if (atom x)
    't
    (if (atom (car x))
        't
        (< (size (rotate x)) (size x))))
```

Use cons/car+cdr twice.

```
(if (atom x)
    't
    (if (atom (car x))
        't
        (< (size (rotate x))
           (size x))))
```

22 Easy, since the focus is in the if else of the premise (atom x).

```
(if (atom x)
    't
    (if (atom (car x))
        't
        (< (size (rotate (cons (car x) (cdr x))))
           (size (cons (car x) (cdr x))))))
```

And again, using the other premise.

```
(if (atom x)
  't
  (if (atom (car x))
    't
    (< (size (rotate (cons (car x)
                           (cdr x))))
       (size (cons (car x)
                   (cdr x))))))
```

23 A snap.

```
(if (atom x)
  't
  (if (atom (car x))
    't
    (< (size (rotate (cons (cons (car (car x))
                                 (cdr (car x)))
                           (cdr x))))
       (size (cons (cons (car (car x))
                         (cdr (car x)))
                   (cdr x))))))
```

Now use the theorem rotate/cons.

```
(if (atom x)
  't
  (if (atom (car x))
    't
    (< (size (rotate (cons (cons (car (car x))
                                 (cdr (car x)))
                           (cdr x))))
       (size (cons (cons (car (car x))
                         (cdr (car x)))
                   (cdr x))))))
```

24 Oh, so that's why we proved rotate/cons!

```
(if (atom x)
  't
  (if (atom (car x))
    't
    (< (size (cons (car (car x))
                   (cons (cdr (car x))
                         (cdr x))))
       (size (cons (cons (car (car x))
                         (cdr (car x)))
                   (cdr x))))))
```

Exactly. Sometimes it's nice to think of a theorem ahead of time that makes it easier to prove things about a function.

25 What if we hadn't thought of rotate/cons ahead of time?

Then this proof would have been a little longer while we worked through all the car/cons and cdr/cons, but we'd get to the same place in the end.

26 That's helpful to know.

Insight: Create Helpers for Repetition

If a proof performs similar sequences of steps over and over, state a theorem that can perform the same rewrite as those steps via the Law of Dethm. Use that theorem in place of the sequence of steps to shorten the proof.

What next?

27 We have to prove that the size of

```
(cons (car (car x))
  (cons (cdr (car x)) (cdr x)))
```

is smaller than the size of

```
(cons
  (cons (car (car x)) (cdr (car x)))
  (cdr x)).
```

Is the size of

```
(cons (car (car x))
  (cons (cdr (car x)) (cdr x)))
```

smaller than the size of

```
(cons
  (cons (car (car x)) (cdr (car x)))
  (cdr x))?
```

28 It doesn't look smaller; it just looks rearranged.

Perhaps (size x) is not the right measure.

29 How can we tell?

Find a counterexample to the totality claim for align.

What value is (size x) equal to when x is '((((french . toast) . and) . maple) . syrup)?

30 '4.

---

31 What value is (size x) equal to when x is (rotate '((((french . toast) . and) . maple) . syrup))?	'4.
32 What is the value of the totality claim for align in frame 20 where x is '((((french . toast) . and) . maple) . syrup)?	'nil, since '4 is not less than '4. Does that mean that align is not a total function?
33 Not necessarily, perhaps we just need to pick a better measure.	How do we do that?
34 The function wt returns '1 for an atom, and twice the wt of the car plus the wt of the cdr for a cons. What value is (wt '((((french . toast) . and) . maple) . syrup)) equal to?	'31. What does the name wt mean?
35 It is short for weight. What value is (wt '(((french . toast) . and) . (maple . syrup))) equal to?	'17.
36 What value is (wt '((french . toast) . (and . (maple . syrup)))) equal to?	'11.

What value is

```
(wt
  '(french . (toast . (and . (maple . syrup)))))
```

equal to?

37 '9.

Define wt.

38 It's quite tricky, isn't it?

Yes, we know. Here it is.

```
(defun wt (x)
  (if (atom x)
      '1
      (+ (+ (wt (car x)) (wt (car x)))
         (wt (cdr x)))))

measure: (size x)
```

39 Is (size x) the *right* measure this time?

Yes, we double-checked. But let's prove it.

```
(if (natp (size x))
    (if (atom x)
        't
        (if (< (size (car x)) (size x))
            (< (size (cdr x)) (size x))
            'nil))
    'nil)
```

40 This is the same step as in frame 20.

```
(if (atom x)
    't
    (if (< (size (car x)) (size x))
        (< (size (cdr x)) (size x))
        'nil))
```

What do we do next?

```
(if (atom x)
    't
    (if (< (size (car x)) (size x))
        (< (size (cdr x)) (size x))
        'nil))
```

41 We use size/car and size/cdr, since we know x is not an atom.

```
(if (atom x)
    't
    (if 't
        't
        'nil))
```

This proof is easy.

```
(if (atom x)
  't
  (if 't
    't
    'nil))
```

42 Indeed, it is.

```
't
```

We still need to prove align is total.

43 This time, we choose wt as the measure.

```
(defun align (x)
  (if (atom x)
    x
    (if (atom (car x))
      (cons (car x) (align (cdr x)))
      (align (rotate x)))))

measure: (wt x)
```

Here is our totality claim, although we could have constructed it step by step.

```
(if (natp (wt x))
  (if (atom x)
    't
    (if (atom (car x))
      (< (wt (cdr x)) (wt x))
      (< (wt (rotate x)) (wt x))))
  'nil)
```

Is (wt x) always a natural number?

44 We don't know, but we suspect so. For now we rewrite (natp (wt x)) to 't and use if-true to drop the outer if.

```
(if (atom x)
  't
  (if (atom (car x))
    (< (wt (cdr x)) (wt x))
    (< (wt (rotate x)) (wt x))))
```

We must prove our claim about wt and natp later.

```
(dethm natp/wt (x)
  (equal (natp (wt x)) 't))
```

Is the definition of wt helpful here?

```
(if (atom x)
  't
  (if (atom (car x))
    (< (wt (cdr x))
       (wt x))
    (< (wt (rotate x)) (wt x))))
```

45 Yes, and so are if-nest-E and the premise (atom x).

```
(if (atom x)
  't
  (if (atom (car x))
    (< (wt (cdr x))
       (+ (+ (wt (car x)) (wt (car x)))
          (wt (cdr x))))
    (< (wt (rotate x)) (wt x))))
```

Now what?

46 We need to know how + and < work.

We assume that $0 + x = x$, $x + y = y + x$, and $x + (y + z) = (x + y) + z$, among other things.

47 Yes, we have heard the rumors.

The Axioms of + and <

```
(dethm identity-+ (x)
  (if (natp x) (equal (+ '0 x) x) 't))
```

```
(dethm commute-+ (x y)
  (equal (+ x y) (+ y x)))
```

```
(dethm associate-+ (x y z)
  (equal (+ (+ x y) z) (+ x (+ y z))))
```

```
(dethm positives-+ (x y)
  (if (< '0 x) (if (< '0 y) (equal (< '0 (+ x y)) 't) 't) 't))
```

```
(dethm natp/+ (x y)
  (if (natp x) (if (natp y) (equal (natp (+ x y)) 't) 't) 't))
```

```
(dethm common-addends-< (x y z)
  (equal (< (+ x z) (+ y z)) (< x y)))
```

Using these new axioms, how can we rewrite our claim?

48 We should cancel the (wt (cdr x)) in both arguments to < using common-addends-<.

```
(if (atom x)
  't
  (if (atom (car x))
    (< (wt (cdr x))
       (+ (+ (wt (car x)) (wt (car x)))
          (wt (cdr x))))
    (< (wt (rotate x)) (wt x))))
```

For that, we must add something to the first argument of <.

49 We can add '0 to the first (wt (cdr x)) without changing its value.

Only if (wt (cdr x)) is a natural number.

```
(if (atom x)
  't
  (if (atom (car x))
    (< (wt (cdr x))
       (+ (+ (wt (car x))
             (wt (car x)))
          (wt (cdr x))))
    (< (wt (rotate x)) (wt x))))
```

50 The claim natp/wt (frame 44) states that it is, and if-true yields a place to put the corresponding premise, as in frame 75 of chapter 9.

```
(if (atom x)
  't
  (if (atom (car x))
    (if (natp (wt (cdr x)))
      (< (wt (cdr x))
         (+ (+ (wt (car x))
               (wt (car x)))
            (wt (cdr x))))
      't)
    (< (wt (rotate x)) (wt x))))
```

Now we have the premise that (wt (cdr x)) is a natural number.

```
(if (atom x)
  't
  (if (atom (car x))
    (if (natp (wt (cdr x)))
      (< (wt (cdr x))
         (+ (+ (wt (car x))
               (wt (car x)))
            (wt (cdr x))))
      't)
    (< (wt (rotate x)) (wt x))))
```

51 Therefore we use identity-+.

```
(if (atom x)
  't
  (if (atom (car x))
    (if (natp (wt (cdr x)))
      (< (+ '0 (wt (cdr x)))
         (+ (+ (wt (car x))
               (wt (car x)))
            (wt (cdr x))))
      't)
    (< (wt (rotate x)) (wt x))))
```

Next?

```
(if (atom x)
  't
  (if (atom (car x))
    (if (natp (wt (cdr x)))
      (< (+ '0
            (wt (cdr x)))
         (+ (+ (wt (car x))
               (wt (car x)))
            (wt (cdr x))))
      't)
    (< (wt (rotate x)) (wt x))))
```

52 At last, we cancel (wt (cdr x)) using common-addends-<.

```
(if (atom x)
  't
  (if (atom (car x))
    (if (natp (wt (cdr x)))
      (< '0
         (+ (wt (car x))
            (wt (car x))))
      't)
    (< (wt (rotate x)) (wt x))))
```

Is (+ (wt (car x)) (wt (car x))) positive?

53 It is positive if (wt (car x)) is positive.

It seems we need another premise. The claim natp/wt rewrites this focus to 't, provided that we eventually prove the claim.

```
(if (atom x)
  't
  (if (atom (car x))
    (if (natp (wt (cdr x)))
      (< '0 (+ (wt (car x))
               (wt (car x))))
      't)
    (< (wt (rotate x)) (wt x))))
```

54 We also make the claim positive/wt, and rewrite this focus to a new premise.

```
(dethm positive/wt (x)
  (equal (< '0 (wt x)) 't))
```

```
(if (atom x)
  't
  (if (atom (car x))
    (if (< '0 (wt (car x)))
      (< '0 (+ (wt (car x))
               (wt (car x))))
      't)
    (< (wt (rotate x)) (wt x))))
```

We must, of course, prove positive/wt.

What do we use the new premise for?

```
(if (atom x)
  't
  (if (atom (car x))
    (if (< '0 (wt (car x)))
      (< '0 (+ (wt (car x))
               (wt (car x))))
      't)
    (< (wt (rotate x)) (wt x))))
```

55 For positives-+, of course., and then followed by if-same.

```
(if (atom x)
  't
  (if (atom (car x))
    't
    (< (wt (rotate x)) (wt x))))
```

One < comparison down.

```
(if (atom x)
  't
  (if (atom (car x))
    't
    (< (wt (rotate x))
       (wt x))))
```

56 And one to go.

```
(if (atom x)
  't
  (if (atom (car x))
    't
    (< (wt (cons (car (car x))
                 (cons (cdr (car x))
                   (cdr x))))
       (wt x))))
```

Here, we use wt. Can we also use atom/cons and if-false to drop the resulting if?

```
(if (atom x)
  't
  (if (atom (car x))
    't
    (< (wt (cons (car (car x))
                 (cons (cdr (car x))
                   (cdr x))))
       (wt x))))
```

57 Yes, and car/cons twice and cdr/cons once as well.

```
(if (atom x)
  't
  (if (atom (car x))
    't
    (< (+ (+ (wt (car (car x)))
             (wt (car (car x))))
          (wt (cons (cdr (car x))
                    (cdr x))))
       (wt x))))
```

Are the same steps useful again?

```
(if (atom x)
  't
  (if (atom (car x))
    't
    (< (+ (+ (wt (car (car x)))
             (wt (car (car x))))
          (wt (cons (cdr (car x))
                    (cdr x))))
       (wt x))))
```

58 Absolutely!

```
(if (atom x)
  't
  (if (atom (car x))
    't
    (< (+ (+ (wt (car (car x)))
             (wt (car (car x))))
          (+ (+ (wt (cdr (car x)))
                (wt (cdr (car x))))
             (wt (cdr x))))
       (wt x))))
```

Do we use the same steps a third time?

59 No, this time we only use wt and if-nest-E, along with the premise (atom x).

```
(if (atom x)
  't
  (if (atom (car x))
    't
    (< (+ (+ (wt (car (car x)))
             (wt (car (car x))))
          (+ (+ (wt (cdr (car x)))
                (wt (cdr (car x))))
             (wt (cdr x))))
       (wt x))))
```

```
(if (atom x)
  't
  (if (atom (car x))
    't
    (< (+ (+ (wt (car (car x)))
             (wt (car (car x))))
          (+ (+ (wt (cdr (car x)))
                (wt (cdr (car x))))
             (wt (cdr x))))
       (+ (+ (wt (car x))
             (wt (car x)))
          (wt (cdr x))))))
```

Let's use wt in these two focuses. Is there an obvious simplification we can do after that?

60 Yes. Since we know that (car x) is not an atom, we use if-nest-E on both focuses.

```
(if (atom x)
  't
  (if (atom (car x))
    't
    (< (+ (+ (wt (car (car x)))
             (wt (car (car x))))
          (+ (+ (wt (cdr (car x)))
                (wt (cdr (car x))))
             (wt (cdr x))))
       (+ (+ (wt (car x))
             (wt (car x)))
          (wt (cdr x))))))
```

```
(if (atom x)
  't
  (if (atom (car x))
    't
    (< (+ (+ (wt (car (car x)))
             (wt (car (car x))))
          (+ (+ (wt (cdr (car x)))
                (wt (cdr (car x))))
             (wt (cdr x))))
       (+ (+ (+ (+ (wt (car (car x)))
                   (wt (car (car x))))
                (wt (cdr (car x))))
             (+ (+ (wt (car (car x)))
                   (wt (car (car x))))
                (wt (cdr (car x)))))
          (wt (cdr x))))))
```

My, how the claim has grown!

61 What next?

Do the new axioms help make our claim smaller?

```
(if (atom x)
  't
  (if (atom (car x))
    't
    (< (+ (+ (wt (car (car x)))
             (wt (car (car x))))
          (+ (+ (wt (cdr (car x)))
                (wt (cdr (car x))))
             (wt (cdr x))))
       (+ (+ (+ (+ (wt (car (car x)))
                   (wt (car (car x))))
                (wt (cdr (car x))))
             (+ (+ (wt (car (car x)))
                   (wt (car (car x))))
                (wt (cdr (car x)))))
          (wt (cdr x))))))
```

62 They might. First we use associate-+.

```
(if (atom x)
  't
  (if (atom (car x))
    't
    (< (+ (+ (+ (wt (car (car x)))
                (wt (car (car x))))
             (+ (wt (cdr (car x)))
                (wt (cdr (car x)))))
          (wt (cdr x)))
       (+ (+ (+ (+ (wt (car (car x)))
                   (wt (car (car x))))
                (wt (cdr (car x))))
             (+ (+ (wt (car (car x)))
                   (wt (car (car x))))
                (wt (cdr (car x)))))
          (wt (cdr x))))))
```

Great choice. Why?

```
(if (atom x)
  't
  (if (atom (car x))
    't
    (< (+ (+ (+ (wt (car (car x)))
                (wt (car (car x))))
             (+ (wt (cdr (car x)))
                (wt (cdr (car x)))))
          (wt (cdr x)))
       (+ (+ (+ (+ (wt (car (car x)))
                   (wt (car (car x))))
                (wt (cdr (car x))))
             (+ (+ (wt (car (car x)))
                   (wt (car (car x))))
                (wt (cdr (car x)))))
          (wt (cdr x))))))
```

63 Now we can cancel out (wt (cdr x)) using common-addends-<.

```
(if (atom x)
  't
  (if (atom (car x))
    't
    (< (+ (+ (wt (car (car x)))
             (wt (car (car x))))
          (+ (wt (cdr (car x)))
             (wt (cdr (car x)))))
       (+ (+ (+ (wt (car (car x)))
                (wt (car (car x))))
             (wt (cdr (car x))))
          (+ (+ (wt (car (car x)))
                (wt (car (car x))))
             (wt (cdr (car x))))))))
```

The claim is already smaller. What next?

```
(if (atom x)
  't
  (if (atom (car x))
    't
    (< (+ (+ (wt (car (car x)))
             (wt (car (car x))))
          (+ (wt (cdr (car x)))
             (wt (cdr (car x)))))
       (+ (+ (+ (wt (car (car x)))
                (wt (car (car x))))
             (wt (cdr (car x))))
          (+ (+ (wt (car (car x)))
                (wt (car (car x))))
             (wt (cdr (car x))))))))
```

64 We prepare to cancel out expressions again using associate-+ and commute-+.

```
(if (atom x)
  't
  (if (atom (car x))
    't
    (< (+ (wt (cdr (car x)))
          (+ (+ (wt (car (car x)))
                (wt (car (car x))))
             (wt (cdr (car x)))))
       (+ (+ (+ (wt (car (car x)))
                (wt (car (car x))))
             (wt (cdr (car x))))
          (+ (+ (wt (car (car x)))
                (wt (car (car x))))
             (wt (cdr (car x))))))))
```

Go on.

```
(if (atom x)
  't
  (if (atom (car x))
    't
    (< (+ (wt (cdr (car x)))
          (+ (+ (wt (car (car x)))
                (wt (car (car x))))
             (wt (cdr (car x)))))
       (+ (+ (+ (wt (car (car x)))
                (wt (car (car x))))
             (wt (cdr (car x))))
          (+ (+ (wt (car (car x)))
                (wt (car (car x))))
             (wt (cdr (car x))))))))
```

65 We use common-addends-< once again to cancel out expressions in this focus.

```
(if (atom x)
  't
  (if (atom (car x))
    't
    (< (wt (cdr (car x)))
       (+ (+ (wt (car (car x)))
             (wt (car (car x))))
          (wt (cdr (car x)))))))
```

Can we cancel out more?

```
(if (atom x)
  't
  (if (atom (car x))
    't
    (< (wt (cdr (car x)))
       (+ (+ (wt (car (car x)))
             (wt (car (car x))))
          (wt (cdr (car x)))))))
```

66 Yes, we can cancel (wt (cdr (car x))) on both sides. First, we use if-true and natp/wt to set up a new premise.

```
(if (atom x)
  't
  (if (atom (car x))
    't
    (if (natp (wt (cdr (car x))))
      (< (wt (cdr (car x)))
         (+ (+ (wt (car (car x)))
               (wt (car (car x))))
            (wt (cdr (car x)))))
      't)))
```

And then?

```
(if (atom x)
  't
  (if (atom (car x))
    't
    (if (natp (wt (cdr (car x))))
      (< (wt (cdr (car x)))
         (+ (+ (wt (car (car x)))
               (wt (car (car x))))
            (wt (cdr (car x)))))
      't)))
```

67 We add '0 to the first argument of <, since we have a premise stating it is a natural number.

```
(if (atom x)
  't
  (if (atom (car x))
    't
    (if (natp (wt (cdr (car x))))
      (< (+ '0 (wt (cdr (car x))))
         (+ (+ (wt (car (car x)))
               (wt (car (car x))))
            (wt (cdr (car x)))))
      't)))
```

Next?

```
(if (atom x)
  't
  (if (atom (car x))
    't
    (if (natp (wt (cdr (car x))))
      (< (+ '0
            (wt (cdr (car x))))
         (+ (+ (wt (car (car x)))
               (wt (car (car x))))
            (wt (cdr (car x)))))
      't)))
```

68 We use common-addends-< one last time.

```
(if (atom x)
  't
  (if (atom (car x))
    't
    (if (natp (wt (cdr (car x))))
      (< '0
         (+ (wt (car (car x)))
            (wt (car (car x)))))
      't)))
```

Now we must show that (+ (wt (car (car x))) (wt (car (car x)))) is positive.

```
(if (atom x)
  't
  (if (atom (car x))
    't
    (if (natp (wt (cdr (car x))))
      (< '0
         (+ (wt (car (car x)))
            (wt (car (car x)))))
      't)))
```

69 To start, we replace the inner if question using natp/wt and positive/wt.

```
(if (atom x)
  't
  (if (atom (car x))
    't
    (if (< '0 (wt (car (car x))))
      (< '0
         (+ (wt (car (car x)))
            (wt (car (car x)))))
      't)))
```

Can we use the new premise?

```
(if (atom x)
  't
  (if (atom (car x))
    't
    (if (< '0 (wt (car (car x))))
      (< '0 (+ (wt (car (car x)))
               (wt (car (car x)))))
      't)))
```

70 We rewrite the remaining < comparison to 't with positives-+.

```
(if (atom x)
  't
  (if (atom (car x))
    't
    (if (< '0 (wt (car (car x))))
      't
      't)))
```

And then?

71 By three uses of if-same, we are done.

We've earned a break.

72 But what about proving natp/wt and positive/wt?

We prove them starting on page 197.

73 Okay, we shall continue, then, but not before taking a small snack.

Now we know that align is total.

74 Can we prove this theorem?

```
(dethm align/align (x)
  (equal (align (align x)) (align x)))
```

That's a good idea. Is align/align likely to be true?

75 In frames 14–17, there are no counterexamples.

Can we prove align/align by induction?

76 The function align is recursive.

Yes, can we use induction based on the definition of align?

77 Probably, but how does that work?

It is like stating the claim that align is total.

78 Yes, we already know how to do that.

Here is our claim.

```
(equal (align (align x)) (align x))
```

Create the inductive claim we must prove. Use Defun Induction on align.

79 Now we have an inductive premise for (align (cdr x)), and another for (align (rotate x)).

```
(if (atom x)
  (equal (align (align x)) (align x))
  (if (atom (car x))
    (if (equal (align (align (cdr x)))
               (align (cdr x)))
      (equal (align (align x)) (align x))
      't)
    (if (equal (align (align (rotate x)))
               (align (rotate x)))
      (equal (align (align x)) (align x))
      't)))
```

80 Use align in both focuses.

The definition of align includes the if question (atom x), which is already a premise. We can use if-nest-A in both focuses, too.

```
(if (atom x)
  (equal (align (align x)) (align x))
  (if (atom (car x))
    (if (equal (align (align (cdr x)))
               (align (cdr x)))
      (equal (align (align x)) (align x))
      't)
    (if (equal (align (align (rotate x)))
               (align (rotate x)))
      (equal (align (align x)) (align x))
      't)))
```

```
(if (atom x)
  (equal (align x) x)
  (if (atom (car x))
    (if (equal (align (align (cdr x)))
               (align (cdr x)))
      (equal (align (align x)) (align x))
      't)
    (if (equal (align (align (rotate x)))
               (align (rotate x)))
      (equal (align (align x)) (align x))
      't)))
```

81 In frame 80, we see that (align x) is equal to x when x is an atom by the definition of align and if-nest-A.

And since x is equal to x, we apply equal-same.

```
(if (atom x)
  (equal (align x) x)
  (if (atom (car x))
    (if (equal (align (align (cdr x)))
               (align (cdr x)))
      (equal (align (align x)) (align x))
      't)
    (if (equal (align (align (rotate x)))
               (align (rotate x)))
      (equal (align (align x)) (align x))
      't)))
```

```
(if (atom x)
  't
  (if (atom (car x))
    (if (equal (align (align (cdr x)))
               (align (cdr x)))
      (equal (align (align x)) (align x))
      't)
    (if (equal (align (align (rotate x)))
               (align (rotate x)))
      (equal (align (align x)) (align x))
      't)))
```

Use align twice, again.

```
(if (atom x)
  't
  (if (atom (car x))
    (if (equal (align (align (cdr x)))
               (align (cdr x)))
      (equal (align
                (align x))
             (align x))
      't)
    (if (equal (align (align (rotate x)))
               (align (rotate x)))
      (equal (align (align x)) (align x))
      't)))
```

82 That's easy. We can simplify it using these premises, as well.

```
(if (atom x)
  't
  (if (atom (car x))
    (if (equal (align (align (cdr x)))
               (align (cdr x)))
      (equal (align
                (cons (car x) (align (cdr x))))
             (cons (car x)
                (align (cdr x))))
      't)
    (if (equal (align (align (rotate x)))
               (align (rotate x)))
      (equal (align (align x)) (align x))
      't)))
```

What else have we used to do this rewrite?

83 We used if-nest-E and if-nest-A for the questions (atom x) and (atom (car x)), respectively, taken from the definition of align.

84

Use the definition of align to rewrite the application whose argument is (cons (car x) (align (cdr x))). Take the next four steps, too.

```
(if (atom x)
  't
  (if (atom (car x))
    (if (equal (align (align (cdr x)))
               (align (cdr x)))
      (equal (align (cons (car x)
                          (align (cdr x))))
             (cons (car x)
               (align (cdr x))))
      't)
    (if (equal (align (align (rotate x)))
               (align (rotate x)))
      (equal (align (align x)) (align x))
      't)))
```

Easy, using atom/cons, cdr/cons, and car/cons twice.

```
(if (atom x)
  't
  (if (atom (car x))
    (if (equal (align (align (cdr x)))
               (align (cdr x)))
      (equal (if 'nil
                 (cons (car x)
                   (align (cdr x)))
                 (if (atom (car x))
                   (cons (car x)
                     (align (align (cdr x))))
                   (align
                     (rotate
                       (cons (car x)
                         (align (cdr x)))))))
             (cons (car x)
               (align (cdr x))))
      't)
    (if (equal (align (align (rotate x)))
               (align (rotate x)))
      (equal (align (align x)) (align x))
      't)))
```

Next simplify the ifs in this focus.

```
(if (atom x)
  't
  (if (atom (car x))
    (if (equal (align (align (cdr x)))
               (align (cdr x)))
      (equal (if 'nil
                 (cons (car x)
                   (align (cdr x)))
                 (if (atom (car x))
                   (cons (car x)
                     (align (align (cdr x))))
                   (align
                     (rotate
                       (cons (car x)
                         (align (cdr x)))))))
             (cons (car x)
               (align (cdr x))))
      't)
    (if (equal (align (align (rotate x)))
               (align (rotate x)))
      (equal (align (align x)) (align x))
      't)))
```

85 Done, using the helpful premise in orange.

```
(if (atom x)
  't
  (if (atom (car x))
    (if (equal (align (align (cdr x)))
               (align (cdr x)))
      (equal (cons (car x)
                   (align (align (cdr x))))
             (cons (car x)
               (align (cdr x))))
      't)
    (if (equal (align (align (rotate x)))
               (align (rotate x)))
      (equal (align (align x)) (align x))
      't)))
```

Apply the inductive premise using equal-if.

```
(if (atom x)
  't
  (if (atom (car x))
    (if (equal (align (align (cdr x)))
               (align (cdr x)))
      (equal (cons (car x)
                   (align (align (cdr x))))
             (cons (car x)
               (align (cdr x))))
      't)
    (if (equal (align (align (rotate x)))
               (align (rotate x)))
      (equal (align (align x)) (align x))
      't)))
```

86 This is the premise corresponding to natural recursion. We can remove the equal in the if answer, too.

```
(if (atom x)
  't
  (if (atom (car x))
    (if (equal (align (align (cdr x)))
               (align (cdr x)))
      't
      't)
    (if (equal (align (align (rotate x)))
               (align (rotate x)))
      (equal (align (align x)) (align x))
      't)))
```

87 Which way do we use the inductive premise? Do we rewrite (align (align (cdr x))) to (align (cdr x)) or vice versa?

Does it matter?

88 No. Either way makes both arguments to equal the same.

Great.

89 What next?

```
(if (atom x)
  't
  (if (atom (car x))
    (if (equal (align (align (cdr x)))
               (align (cdr x)))
      't
      't)
    (if (equal (align (align (rotate x)))
               (align (rotate x)))
      (equal (align (align x)) (align x))
      't)))
```

We drop an if.

```
(if (atom x)
  't
  (if (atom (car x))
    't
    (if (equal (align (align (rotate x)))
               (align (rotate x)))
      (equal (align (align x)) (align x))
      't)))
```

90 Once again, expand (align x).

```
(if (atom x)
  't
  (if (atom (car x))
    't
    (if (equal (align (align (rotate x)))
               (align (rotate x)))
      (equal (align (align x))
             (align x))
      't)))
```

And four if-nest-Es using two premises.

```
(if (atom x)
  't
  (if (atom (car x))
    't
    (if (equal (align (align (rotate x)))
               (align (rotate x)))
      (equal (align (align (rotate x)))
             (align (rotate x)))
      't)))
```

There's only one more case. We have to use the other inductive premise.

```
(if (atom x)
  't
  (if (atom (car x))
    't
    (if (equal (align (align (rotate x)))
               (align (rotate x)))
      (equal (align (align (rotate x)))
             (align (rotate x)))
      't)))
```

91 Like this?

```
(if (atom x)
  't
  (if (atom (car x))
    't
    (if (equal (align (align (rotate x)))
               (align (rotate x)))
      (equal (align (rotate x))
             (align (rotate x)))
      't)))
```

Correct.

```
(if (atom x)
  't
  (if (atom (car x))
    't
    (if (equal (align (align (rotate x)))
               (align (rotate x)))
      (equal (align (rotate x))
             (align (rotate x)))
      't)))
```

92 And now we are done.

```
(if (atom x)
  't
  (if (atom (car x))
    't
    't))
```

That's it for this proof. Feel free to visit J-Bob on page 196.

93 Finally, Q.E.D.!

Quick, Eat Doughnuts!

RECESS

1 We're now going to use J-Bob to explore how we rewrite expressions.

What is J-Bob?[†]

[†]Thank you, J Moore and Bob Boyer.

---

2 Patience. We will discover what J-Bob is by experimenting with it.[†]

[†]To play along, find J-Bob on page 202 or at `http://the-little-prover.org/`.

Intriguing.

---

3 What value is this expression equal to?

```
(J-Bob/step (prelude)
  '(car (cons 'ham '(cheese)))
  '())
```

The value,

```
'(car (cons 'ham '(cheese)))
```

which represents the expression we start with in frame 8 of chapter 1.

---

4 What are `J-Bob/step`'s three arguments?

Here are our words:
The first argument to `J-Bob/step` is a list of representations of definitions, in this case `(prelude)` representing J-Bob's axioms and initial functions. The second argument represents an expression to rewrite. The third argument is a list of steps, processed first to last, to rewrite the expression. Here the list of steps is empty.

---

5 Do the first and second arguments to `J-Bob/step` make sense?

What are representations of expressions and definitions?

6 We represent expressions and definitions as quoted values. For instance, to represent the expression `x`, we write `'x`, which is short for `(quote x)`. Our expressions include variable names, e.g. `'x`; quoted values e.g. `''eggs`; `if` expressions, e.g. `'(if x y z)`, described in chapter 2; and function applications e.g. `'(cons 'eggs x)`. Definitions may be theorems or functions, such as `'(dethm truth () 't)` and `'(defun id (x) x)`, described in chapter 3.

And what are steps to rewrite an expression?

---

7 We shall see. What value is this expression equal to?

```
(J-Bob/step (prelude)
  '(car (cons 'ham '(cheese)))
  '((() (car/cons 'ham '(cheese)))))
```

The value,

```
''ham
```

which represents the result of the rewrite in frame 8 of chapter 1. But why is there an empty list in the first step?

---

8 That list is the *path* to the focus. The path is a list of directions from the current expression to the subexpression representing the focus in the pending rewrite. Here, since the focus is the entire expression (otherwise put, the context is empty), the path is empty as well.

Okay.

---

9 What value is this expression equal to?

```
(J-Bob/step (prelude)
  '(equal 'flapjack (atom (cons a b)))
  '(((2) (atom/cons a b))
    (() (equal 'flapjack 'nil))))
```

The value,

```
''nil
```

which represents the result of the rewrite in frame 16 of chapter 1.

What value is this expression equal to?

```
(J-Bob/step (prelude)
  '(atom (cdr (cons (car (cons p q)) '())))
  '(((1 1 1) (car/cons p q))
    ((1) (cdr/cons p '()))
    (() (atom '()))))
```

10 The value,

```
''t
```

which represents the result of the rewrite in frame 29 of chapter 1. But how does J-Bob find this value?

How do we find the step in frame 29 of chapter 1?

11 We use `car/cons` on `p` and `q` in the focus.

Does J-Bob have that information?

12 Yes, we see `cons` with arguments `p` and `q`. But what about the focus?

Good question. Where is the focus in frame 29 of chapter 1?

13 Inside the 1st argument to `atom`, which is `(cdr (cons (car (cons p q)) '()))`; inside the 1st argument to `cdr`, which is `(cons (car (cons p q)) '())`, which is inside 1st argument to `cons`, which is `(car (cons p q))`, which then allows us to use `car/cons` with `p` and `q`.

What value is this expression equal to?

```
(J-Bob/step (prelude)
  '(if a c c)
  '())
```

14 The value,

```
'(if a c c)
```

which represents the expression we start with in frame 5 of chapter 2.

What value is this expression equal to?

```
(J-Bob/step (prelude)
  '(if a c c)
  '((() (if-same a c))))
```

15 The value,

```
'c
```

which represents the result of the rewrite in frame 5 of chapter 2.

---

16 What value is this expression equal to?

```
(J-Bob/step (prelude)
  '(if a c c)
  '((() (if-same a c))
    (()
     (if-same
       (if (equal a 't)
           (if (equal 'nil 'nil) a b)
           (equal 'or
                  (cons 'black '(coffee))))
       c))))
```

The value,

```
'(if (if (equal a 't)
         (if (equal 'nil 'nil)
             a
             b)
         (equal 'or
                (cons 'black '(coffee))))
     c
     c)
```

which represents the result of the rewrite in frame 7 of chapter 2.

---

17 What value is this expression equal to?

```
(J-Bob/step (prelude)
  '(if a c c)
  '((() (if-same a c))
    (()
     (if-same
       (if (equal a 't)
           (if (equal 'nil 'nil)
               a
               b)
           (equal 'or
                  (cons 'black '(coffee))))
       c))
    ((Q E 2) (cons 'black '(coffee)))))
```

The value,

```
'(if (if (equal a 't)
         (if (equal 'nil 'nil)
             a
             b)
         (equal 'or
                '(black coffee)))
     c
     c)
```

which represents the result of the rewrite in frame 7 of chapter 2. But how does J-Bob find this value?

---

18 How do we find the step in frame 7 of chapter 2?

We use cons on 'black and '(coffee) in the focus.

---

19 Does J-Bob have that information?

Yes, we see `cons` with arguments `'black` and `'(coffee)`. But what about the focus?

---

20 An excellent question. Where is the focus in frame 11 of chapter 2?

Inside the question of the outer if, in the else of the middle if, in the second argument to the outer cons.

21

Exactly. The `(Q E 2)` in the proof step in frame 17 of chapter 2 describes the path to this focus: in the Question of the outer `if`, in the Else of the middle `if`, and in the 2nd argument of the outer `cons`. How would we find the focus in the *first* argument of the outer `cons`?

`(Q E 1)`, right?

22

Yes. What value is this expression equal to?

```
(J-Bob/step (prelude)
  '(if a c c)
  '((() (if-same a c))
    (()
     (if-same
       (if (equal a 't)
           (if (equal 'nil 'nil)
               a
               b)
           (equal 'or
                  (cons 'black '(coffee))))
       c))
    ((Q E 2) (cons 'black '(coffee)))
    ((Q A Q) (equal-same 'nil))))
```

As expected, the result of the rewrite in frame 12 of chapter 2.

```
'(if (if (equal a 't)
         (if 't
             a
             b)
         (equal 'or
                '(black coffee)))
     c
     c)
```

23

One more time.

```
(J-Bob/step (prelude)
  '(if a c c)
  '((() (if-same a c))
    (()
     (if-same
       (if (equal a 't)
           (if (equal 'nil 'nil)
               a
               b)
           (equal 'or
                  (cons 'black '(coffee))))
       c))
    ((Q E 2) (cons 'black '(coffee)))
    ((Q A Q) (equal-same 'nil))
    ((Q A 2) (if-true a b))))
```

We expected the result of the rewrite in frame 14 of chapter 2, but instead the expression has not changed from frame 22 of chapter 2. Why?

```
'(if (if (equal a 't)
         (if 't a b)
         (equal 'or
                '(black coffee)))
     c
     c)
```

What is the path to this focus in frame 14 of chapter 2?

24 This focus is in the question of the outer `if` and in the else of the middle `if`.

And what path do we give `J-Bob/step` for the last step in frame 23 of chapter 2?

25 `(Q E 2)`, which means in the `Q` of the outer `if`, in the `E` of the middle `if`, and in the 2nd argument of an application. But there is no application there!

Exactly. J-Bob stops when a step is incorrect, and does not perform any more rewrites. The `J-Bob/step` function returns whatever expression it has when it encounters the invalid step.

26 That is quite useful to know.

Just once more; this time for certain. Now we have the correct path for the rewrite.

```
(J-Bob/step (prelude)
  '(if a c c)
  '((() (if-same a c))
    (()
     (if-same
       (if (equal a 't)
           (if (equal 'nil 'nil)
               a
               b)
           (equal 'or
                  (cons 'black '(coffee))))
       c))
    ((Q E 2) (cons 'black '(coffee)))
    ((Q A Q) (equal-same 'nil))
    ((Q A) (if-true a b))))
```

27 And this time, we get the right result.

```
'(if (if (equal a 't)
         a
         (equal 'or
                '(black coffee)))
     c
     c)
```

How does J-Bob arrive at this step?

28 The step `((Q A) (if-true a b))` describes the focus in the question of the outer `if` and the answer of the middle `if` as `(Q A)`. The axiom to use is `if-true` and its arguments are `a` and `b`.

29

What value is this expression equal to?

```
(J-Bob/prove (prelude)
  '())
```

A representation of the expression `'t`.

```
''t
```

30

What value is this expression equal to?

```
(J-Bob/prove (prelude)
  '(((defun pair (x y)
       (cons x (cons y '())))
     nil)))
```

Also `''t`. But why? What are the two arguments to `J-Bob/prove`?

```
''t
```

31

The first argument is a list of definitions, and the second argument is a list of *proof attempts*. Each proof attempt starts with a definition and a *seed*. A seed is extra information outside of the definition used to generate the claim that must be proved. For the kinds of proofs we see in chapter 3, the seed is always `nil`. In later chapters, we see other kinds of seeds.

Why does `J-Bob/prove` return `''t`?

32

The result of `J-Bob/prove` in frame 30 of chapter 3 is `''t` because the given `defun` is non-recursive; the only proof attempt therefore succeeds.

All right.

33

What value is this expression equal to?

```
(J-Bob/prove (prelude)
  '(((defun pair (x y)
       (cons x (cons y '())))
     nil)
    ((defun first-of (x)
       (car x))
     nil)
    ((defun second-of (x)
       (car (cdr x)))
     nil)))
```

Still `''t`, because `first-of` and `second-of` are non-recursive.

```
''t
```

What value is this expression equal to?

```
(J-Bob/prove (prelude)
  '(((defun pair (x y)
       (cons x (cons y '())))
     nil)
    ((defun first-of (x)
       (car x))
     nil)
    ((defun second-of (x)
       (car (cdr x)))
     nil)
    ((dethm first-of-pair (a b)
       (equal (first-of (pair a b)) a))
     nil)))
```

34 The body of `first-of-pair`, because we have not proved the theorem.

```
'(equal (first-of (pair a b)) a)
```

What value is this expression equal to?

```
(J-Bob/prove (prelude)
  '(((defun pair (x y)
       (cons x (cons y '())))
     nil)
    ((defun first-of (x)
       (car x))
     nil)
    ((defun second-of (x)
       (car (cdr x)))
     nil)
    ((dethm first-of-pair (a b)
       (equal (first-of (pair a b)) a))
     nil
     ((1 1) (pair a b)))))
```

35 It is a representation of the result of the rewrite in frame 12 of chapter 3.

```
'(equal (first-of (cons a (cons b '()))) a)
```

And why?

36 Perhaps because we have one step in the proof attempt for `first-of-pair`.

Precisely. As we add steps to a proof, J-Bob performs the rewrites in order.

37 Handy.

What value is this expression equal to?

```
(J-Bob/prove (prelude)
  '(((defun pair (x y)
       (cons x (cons y '())))
     nil)
    ((defun first-of (x)
       (car x))
     nil)
    ((defun second-of (x)
       (car (cdr x)))
     nil)
    ((dethm first-of-pair (a b)
       (equal (first-of (pair a b)) a))
     nil
     ((1 1) (pair a b))
     ((1) (first-of (cons a (cons b '()))))
     ((1) (car/cons a (cons b '())))
     (() (equal-same a)))))
```

38 ''t again, because the second argument to J-Bob/prove includes a complete proof of first-of-pair.

```
''t
```

Now we have proved a theorem with J-Bob.

39 What if we want to prove more?

We can add more proofs to the second argument of J-Bob/prove.

40 Does the second argument keep growing forever?

Perhaps. But we can also use J-Bob/define if we want to record our proofs so far.

41 How does that work?

42 We replace J-Bob/prove with J-Bob/define, which produces a list of all the theorems we know or have proved so far. We put this expression in a defun, here named prelude+first-of-pair, to record our work.

```
(defun prelude+first-of-pair ()
  (J-Bob/define (prelude)
    '(((defun pair (x y)
         (cons x (cons y '())))
       nil)
      ((defun first-of (x)
         (car x))
       nil)
      ((defun second-of (x)
         (car (cdr x)))
       nil)
      ((dethm first-of-pair (a b)
         (equal (first-of (pair a b)) a))
       nil
       ((1 1) (pair a b))
       ((1) (first-of (cons a (cons b '()))))
       ((1) (car/cons a (cons b '())))
       (() (equal-same a))))))
```

Then what?

---

43 Once we record our work with J-Bob/define, we can use it to start a new proof attempt. For example, what value is this expression equal to?

```
(J-Bob/prove (prelude)
  '(((dethm second-of-pair (a b)
       (equal (second-of (pair a b)) b))
     nil)))
```

That is not what we expect. What is wrong?

```
''nil
```

---

44 Can we prove second-of-pair based only on the prelude?

Of course not. We need the definitions of second-of and pair that are saved in prelude+first-of-pair to prove second-of-pair.

In order to use the definitions in `prelude+first-of-pair`, we must pass them as the first argument to `J-Bob/prove`.

```
(J-Bob/prove (prelude+first-of-pair)
  '(((dethm second-of-pair (a b)
       (equal (second-of (pair a b)) b))
     nil)))
```

45 And now we start a proof attempt for `second-of-pair`.

```
'(equal (second-of (pair a b)) b)
```

What if we give several proof attempts to `J-Bob/prove` without finishing them?

```
(J-Bob/prove (prelude+first-of-pair)
  '(((dethm second-of-pair (a b)
       (equal (second-of (pair a b)) b))
     nil)
    ((defun in-pair? (xs)
       (if (equal (first-of xs) '?)
           't
           (equal (second-of xs) '?)))
     nil)
    ((dethm in-first-of-pair (b)
       (equal (in-pair? (pair '? b)) 't))
     nil)
    ((dethm in-second-of-pair (a)
       (equal (in-pair? (pair a '?)) 't))
     nil)))
```

46 Does J-Bob ignore the first two `dethm`s?

```
'(equal (in-pair? (pair a '?)) 't)
```

`J-Bob/prove` shows us the expression from the *last* unfinished proof attempt, but we must finish *all* the proofs to get `''t`.

47 That makes sense. Does `J-Bob/define` start at the last proof, too?

`J-Bob/define` only includes a definition in its result when it has been completely proved, including everything that comes before it. Use `J-Bob/prove` to work on unfinished proofs, then use `J-Bob/define` when the proofs are finished.

48 Got it.

49 Below, we include complete proofs that are certified by J-Bob.

We can't wait to play along.

50 What value is this expression equal to?

```
(J-Bob/prove (prelude)
  '(((defun list? (x)
       (if (atom x)
           (equal x '())
           (list? (cdr x))))
     nil)))
```

A representation of the expression `'nil`. Is that because `list?` is recursive?

```
''nil
```

51 Exactly. J-Bob won't give us `''t` until we prove each `defun` total.

How do we do that?

52 We must pass the measure of `list?`, defined in frame 69 of chapter 4, as the seed of our proof attempt. What does `J-Bob/prove` produce?

```
(J-Bob/prove (prelude)
  '(((defun list? (x)
       (if (atom x)
           (equal x '())
           (list? (cdr x))))
     (size x))))
```

A representation of the totality claim for `list?`.

```
'(if (natp (size x))
     (if (atom x)
         't
         (< (size (cdr x)) (size x)))
     'nil)
```

We prove a totality claim from a `defun` by adding proof steps, just like any other claim.

```
(J-Bob/prove (prelude)
  '(((defun list? (x)
       (if (atom x)
           (equal x '())
           (list? (cdr x))))
     (size x)
     ((Q) (natp/size x))
     (()
      (if-true
        (if (atom x)
            't
            (< (size (cdr x)) (size x)))
        'nil))
     ((E) (size/cdr x))
     (() (if-same (atom x) 't)))))
```

53 Sensible.

```
''t
```

Can we prove totality of `memb?` and `remb` using J-Bob?

54 Perhaps. Is that difficult?

Not at all. We use the measure expressions of `memb?` and `remb` as the seed for their respective proof attempts.

```
(J-Bob/prove (prelude)
  '(((defun memb? (xs)
       (if (atom xs)
           'nil
           (if (equal (car xs) '?)
               't
               (memb? (cdr xs)))))
     (size xs))
    ((defun remb (xs)
       (if (atom xs)
           '()
           (if (equal (car xs) '?)
               (remb (cdr xs))
               (cons (car xs) (remb (cdr xs))))))
     (size xs))))
```

55 But it appears we must still fill in the proof steps.

```
'(if (natp (size xs))
     (if (atom xs)
         't
         (< (size (cdr xs)) (size xs)))
     'nil)
```

Naturally.

56 Do we need to start from `prelude+first-of-pair` instead of `prelude`?

We can use `prelude+first-of-pair`, or we can use `prelude` if we do not need the definitions of `pair`, `first-of`, and `first-of-pair`.

57 And can we save our work when we have proved `memb?` and `remb` are total?

Certainly. We can save the definitions as `prelude+memb?+remb` with `J-Bob/define`.

58 Good.

Are we ready to try induction with J-Bob?

59 Ready and eager.

Saved our work from frame 56 of chapter 5?

60 Of course.

Then we must use `list-induction` in the seed of our proof attempt and give a single variable name for the list we want to consider for induction. We should also remember to start off with `memb?` and `remb` from `chapter5`.

```
(J-Bob/prove (chapter5)
  '(((dethm memb?/remb (xs)
       (equal (memb? (remb xs)) 'nil))
     (list-induction xs))))
```

61 And then?

```
'(if (atom xs)
     (equal (memb? (remb xs)) 'nil)
     (if (equal (memb? (remb (cdr xs))) 'nil)
         (equal (memb? (remb xs)) 'nil)
         't))
```

After we generate the inductive claim, we just add steps to the proof until it is done.

62 That's not hard at all.

63 Ready to do more induction with J-Bob?

We are star-struck.

64 Star Induction works similarly to List Induction. Use `star-induction` in the seed for a `dethm` and name the variable on which to do induction.

```
(J-Bob/prove (prelude)
  '(((defun sub (x y)
       (if (atom y)
           (if (equal y '?) x y)
           (cons (sub x (car y))
             (sub x (cdr y)))))
     (size y))
    ((defun ctx? (x)
       (if (atom x)
           (equal x '?)
           (if (ctx? (car x))
               't
               (ctx? (cdr x)))))
     (size x))
    ((dethm ctx?/sub (x y)
       (if (ctx? x)
           (if (ctx? y)
               (equal (ctx? (sub x y)) 't)
               't)
           't))
     (star-induction y))))
```

How excellently simple.

```
'(if (atom y)
     (if (ctx? x)
         (if (ctx? y)
             (equal (ctx? (sub x y)) 't)
             't)
         't)
     (if (if (ctx? x)
             (if (ctx? (car y))
                 (equal
                   (ctx? (sub x (car y)))
                   't)
                 't)
             't)
         (if (if (ctx? x)
                 (if (ctx? (cdr y))
                     (equal
                       (ctx? (sub x (cdr y)))
                       't)
                     't)
                 't)
             (if (ctx? x)
                 (if (ctx? y)
                     (equal
                       (ctx? (sub x y))
                       't)
                     't)
                 't)
             't)
         't))
```

65 Do we already know how to do Defun Induction with J-Bob?

Yes, we do it in frames 61 and 64. The functions `list-induction` and `star-induction` are defined in `prelude` on page 214.

The Proof of the Pudding

Chapter 1 Examples

```
(defun chapter1.example1 ()
  (J-Bob/step (prelude)
    '(car (cons 'ham '(eggs)))
    '(((1) (cons 'ham '(eggs)))
      (() (car '(ham eggs))))))

(defun chapter1.example2 ()
  (J-Bob/step (prelude)
    '(atom '())
    '((() (atom '())))))

(defun chapter1.example3 ()
  (J-Bob/step (prelude)
    '(atom (cons 'ham '(eggs)))
    '(((1) (cons 'ham '(eggs)))
      (() (atom '(ham eggs))))))

(defun chapter1.example4 ()
  (J-Bob/step (prelude)
    '(atom (cons a b))
    '((() (atom/cons a b)))))

(defun chapter1.example5 ()
  (J-Bob/step (prelude)
    '(equal 'flapjack (atom (cons a b)))
    '(((2) (atom/cons a b))
      (() (equal 'flapjack 'nil)))))

(defun chapter1.example6 ()
  (J-Bob/step (prelude)
    '(atom (cdr (cons (car (cons p q)) '())))
    '(((1 1 1) (car/cons p q))
      ((1) (cdr/cons p '()))
      (() (atom '())))))

(defun chapter1.example7 ()
  (J-Bob/step (prelude)
    '(atom (cdr (cons (car (cons p q)) '())))
    '(((1) (cdr/cons (car (cons p q)) '()))
      (() (atom '())))))

(defun chapter1.example8 ()
  (J-Bob/step (prelude)
    '(car (cons (equal (cons x y) (cons x y)) '(and crumpets)))
    '(((1 1) (equal-same (cons x y)))
      ((1) (cons 't '(and crumpets)))
      (() (car '(t and crumpets))))))

(defun chapter1.example9 ()
  (J-Bob/step (prelude)
    '(equal (cons x y) (cons 'bagels '(and lox)))
    '((() (equal-swap (cons x y) (cons 'bagels '(and lox)))))))

(defun chapter1.example10 ()
  (J-Bob/step (prelude)
    '(cons y (equal (car (cons (cdr x) (car y))) (equal (atom x) 'nil)))
    '(((2 1) (car/cons (cdr x) (car y))))))
```

```
(defun chapter1.example11 ()
  (J-Bob/step (prelude)
    '(cons y (equal (car (cons (cdr x) (car y))) (equal (atom x) 'nil)))
    '(((2 1) (car/cons (car (cons (cdr x) (car y))) '(oats)))
      ((2 2 2) (atom/cons (atom (cdr (cons a b))) (equal (cons a b) c)))
      ((2 2 2 1 1 1) (cdr/cons a b))
      ((2 2 2 1 2) (equal-swap (cons a b) c)))))

(defun chapter1.example12 ()
  (J-Bob/step (prelude)
    '(atom (car (cons (car a) (cdr b))))
    '(((1) (car/cons (car a) (cdr b))))))
```

Chapter 2 Examples

```
(defun chapter2.example1 ()
  (J-Bob/step (prelude)
    '(if (car (cons a b)) c c)
    '(((Q) (car/cons a b))
      (() (if-same a c))
      (()
       (if-same
         (if (equal a 't) (if (equal 'nil 'nil) a b) (equal 'or (cons 'black '(coffee))))
         c))
      ((Q E 2) (cons 'black '(coffee)))
      ((Q A Q) (equal-same 'nil))
      ((Q A) (if-true a b))
      ((Q A) (equal-if a 't)))))

(defun chapter2.example2 ()
  (J-Bob/step (prelude)
    '(if (atom (car a))
         (if (equal (car a) (cdr a)) 'hominy 'grits)
         (if (equal (cdr (car a)) '(hash browns))
             (cons 'ketchup (car a))
             (cons 'mustard (car a))))
    '(((E A 2) (cons/car+cdr (car a)))
      ((E A 2 2) (equal-if (cdr (car a)) '(hash browns))))))

(defun chapter2.example3 ()
  (J-Bob/step (prelude)
    '(cons 'statement
       (cons (if (equal a 'question) (cons n '(answer)) (cons n '(else)))
         (if (equal a 'question) (cons n '(other answer)) (cons n '(other else)))))
    '(((2)
       (if-same (equal a 'question)
         (cons (if (equal a 'question) (cons n '(answer)) (cons n '(else)))
           (if (equal a 'question) (cons n '(other answer)) (cons n '(other else))))))
      ((2 A 1) (if-nest-A (equal a 'question) (cons n '(answer)) (cons n '(else))))
      ((2 E 1) (if-nest-E (equal a 'question) (cons n '(answer)) (cons n '(else))))
      ((2 A 2)
       (if-nest-A (equal a 'question) (cons n '(other answer)) (cons n '(other else))))
      ((2 E 2)
       (if-nest-E (equal a 'question)
         (cons n '(other answer))
         (cons n '(other else)))))))
```

Chapter 3 Proofs

```
(defun defun.pair ()
  (J-Bob/define (prelude)
    '(((defun pair (x y)
         (cons x (cons y '())))
       nil))))

(defun defun.first-of ()
  (J-Bob/define (defun.pair)
    '(((defun first-of (x)
         (car x))
       nil))))

(defun defun.second-of ()
  (J-Bob/define (defun.first-of)
    '(((defun second-of (x)
         (car (cdr x)))
       nil))))

(defun dethm.first-of-pair ()
  (J-Bob/define (defun.second-of)
    '(((dethm first-of-pair (a b)
         (equal (first-of (pair a b)) a))
       nil
       ((1 1) (pair a b))
       ((1) (first-of (cons a (cons b '()))))
       ((1) (car/cons a (cons b '())))
       (() (equal-same a))))))

(defun dethm.second-of-pair ()
  (J-Bob/define (dethm.first-of-pair)
    '(((dethm second-of-pair (a b)
         (equal (second-of (pair a b)) b))
       nil
       ((1) (second-of (pair a b)))
       ((1 1 1) (pair a b))
       ((1 1) (cdr/cons a (cons b '())))
       ((1) (car/cons b '()))
       (() (equal-same b))))))

(defun defun.in-pair? ()
  (J-Bob/define (dethm.second-of-pair)
    '(((defun in-pair? (xs)
         (if (equal (first-of xs) '?) 't (equal (second-of xs) '?)))
       nil))))

(defun dethm.in-first-of-pair ()
  (J-Bob/define (defun.in-pair?)
    '(((dethm in-first-of-pair (b)
         (equal (in-pair? (pair '? b)) 't))
       nil
       ((1 1) (pair '? b))
       ((1) (in-pair? (cons '? (cons b '()))))
       ((1 Q 1) (first-of (cons '? (cons b '()))))
       ((1 Q 1) (car/cons '? (cons b '())))
       ((1 Q) (equal-same '?))
       ((1) (if-true 't (equal (second-of (cons '? (cons b '()))) '?)))
       (() (equal-same 't))))))
```

```
(defun dethm.in-second-of-pair ()
  (J-Bob/define (dethm.in-first-of-pair)
    '(((dethm in-second-of-pair (a)
         (equal (in-pair? (pair a '?)) 't))
       nil
       ((1 1) (pair a '?))
       ((1) (in-pair? (cons a (cons '? '()))))
       ((1 Q 1) (first-of (cons a (cons '? '()))))
       ((1 Q 1) (car/cons a (cons '? '())))
       ((1 E 1) (second-of (cons a (cons '? '()))))
       ((1 E 1 1) (cdr/cons a (cons '? '())))
       ((1 E 1) (car/cons '? '()))
       ((1 E) (equal-same '?))
       ((1) (if-same (equal a '?) 't))
       (() (equal-same 't))))))
```

Chapter 4 Proofs

```
(defun defun.list0? ()
  (J-Bob/define (dethm.in-second-of-pair)
    '(((defun list0? (x)
         (equal x '()))
       nil))))

(defun defun.list1? ()
  (J-Bob/define (defun.list0?)
    '(((defun list1? (x)
         (if (atom x) 'nil (list0? (cdr x))))
       nil))))

(defun defun.list2? ()
  (J-Bob/define (defun.list1?)
    '(((defun list2? (x)
         (if (atom x) 'nil (list1? (cdr x))))
       nil))))

(defun dethm.contradiction ()
  (J-Bob/prove
    (list-extend (prelude)
      '(defun partial (x)
         (if (partial x) 'nil 't)))
    '(((dethm contradiction () 'nil)
       nil
       (() (if-same (partial x) 'nil))
       ((A) (if-nest-A (partial x) 'nil 't))
       ((E) (if-nest-E (partial x) 't 'nil))
       ((A Q) (partial x))
       ((E Q) (partial x))
       ((A Q) (if-nest-A (partial x) 'nil 't))
       ((E Q) (if-nest-E (partial x) 'nil 't))
       ((A) (if-false 'nil 't))
       ((E) (if-true 't 'nil))
       (() (if-same (partial x) 't))))))
```

```
(defun defun.list? ()
  (J-Bob/define (defun.list2?)
    '(((defun list? (x)
         (if (atom x) (equal x '()) (list? (cdr x))))
       (size x)
       ((Q) (natp/size x))
       (() (if-true (if (atom x) 't (< (size (cdr x)) (size x))) 'nil))
       ((E) (size/cdr x))
       (() (if-same (atom x) 't))))))

(defun defun.sub ()
  (J-Bob/define (defun.list?)
    '(((defun sub (x y)
         (if (atom y) (if (equal y '?) x y) (cons (sub x (car y)) (sub x (cdr y)))))
       (size y)
       ((Q) (natp/size y))
       (()
        (if-true
          (if (atom y)
              't
              (if (< (size (car y)) (size y)) (< (size (cdr y)) (size y)) 'nil))
          'nil))
       ((E Q) (size/car y))
       ((E A) (size/cdr y))
       ((E) (if-true 't 'nil))
       (() (if-same (atom y) 't))))))
```

Chapter 5 Proofs

```
(defun defun.memb? ()
  (J-Bob/define (defun.sub)
    '(((defun memb? (xs)
         (if (atom xs) 'nil (if (equal (car xs) '?) 't (memb? (cdr xs)))))
       (size xs)
       ((Q) (natp/size xs))
       (()
        (if-true
          (if (atom xs) 't (if (equal (car xs) '?) 't (< (size (cdr xs)) (size xs))))
          'nil))
       ((E E) (size/cdr xs))
       ((E) (if-same (equal (car xs) '?) 't))
       (() (if-same (atom xs) 't))))))

(defun defun.remb ()
  (J-Bob/define (defun.memb?)
    '(((defun remb (xs)
         (if (atom xs)
             '()
             (if (equal (car xs) '?) (remb (cdr xs)) (cons (car xs) (remb (cdr xs))))))
       (size xs)
       ((Q) (natp/size xs))
       (() (if-true (if (atom xs) 't (< (size (cdr xs)) (size xs))) 'nil))
       ((E) (size/cdr xs))
       (() (if-same (atom xs) 't))))))
```

```
(defun dethm.memb?/remb0 ()
  (J-Bob/define (defun.remb)
    '(((dethm memb?/remb0 ()
         (equal (memb? (remb '())) 'nil))
       nil
       ((1 1) (remb '()))
       ((1 1 Q) (atom '()))
       ((1 1)
        (if-true '()
          (if (equal (car '()) '?) (remb (cdr '())) (cons (car '()) (remb (cdr '()))))))
       ((1) (memb? '()))
       ((1 Q) (atom '()))
       ((1) (if-true 'nil (if (equal (car '()) '?) 't (memb? (cdr '())))))
       (() (equal-same 'nil))))))

(defun dethm.memb?/remb1 ()
  (J-Bob/define (dethm.memb?/remb0)
    '(((dethm memb?/remb1 (x1)
         (equal (memb? (remb (cons x1 '()))) 'nil))
       nil
       ((1 1) (remb (cons x1 '())))
       ((1 1 Q) (atom/cons x1 '()))
       ((1 1)
        (if-false '()
          (if (equal (car (cons x1 '())) '?)
              (remb (cdr (cons x1 '())))
              (cons (car (cons x1 '())) (remb (cdr (cons x1 '())))))))
       ((1 1 Q 1) (car/cons x1 '()))
       ((1 1 A 1) (cdr/cons x1 '()))
       ((1 1 E 1) (car/cons x1 '()))
       ((1 1 E 2 1) (cdr/cons x1 '()))
       ((1)
        (if-same (equal x1 '?)
          (memb? (if (equal x1 '?) (remb '()) (cons x1 (remb '()))))))
       ((1 A 1) (if-nest-A (equal x1 '?) (remb '()) (cons x1 (remb '()))))
       ((1 E 1) (if-nest-E (equal x1 '?) (remb '()) (cons x1 (remb '()))))
       ((1 A) (memb?/remb0))
       ((1 E) (memb? (cons x1 (remb '()))))
       ((1 E Q) (atom/cons x1 (remb '())))
       ((1 E)
        (if-false 'nil
          (if (equal (car (cons x1 (remb '()))) '?)
              't
              (memb? (cdr (cons x1 (remb '())))))))
       ((1 E Q 1) (car/cons x1 (remb '())))
       ((1 E E 1) (cdr/cons x1 (remb '())))
       ((1 E) (if-nest-E (equal x1 '?) 't (memb? (remb '()))))
       ((1 E) (memb?/remb0))
       ((1) (if-same (equal x1 '?) 'nil))
       (() (equal-same 'nil))))))

(defun dethm.memb?/remb2 ()
  (J-Bob/define (dethm.memb?/remb1)
    '(((dethm memb?/remb2 (x1 x2)
         (equal (memb? (remb (cons x2 (cons x1 '())))) 'nil))
       nil
```

```
((1 1) (remb (cons x2 (cons x1 '()))))
((1 1 Q) (atom/cons x2 (cons x1 '())))
((1 1)
 (if-false '()
   (if (equal (car (cons x2 (cons x1 '()))) '?)
       (remb (cdr (cons x2 (cons x1 '()))))
       (cons (car (cons x2 (cons x1 '())))
         (remb (cdr (cons x2 (cons x1 '())))))))))
((1 1 Q 1) (car/cons x2 (cons x1 '())))
((1 1 A 1) (cdr/cons x2 (cons x1 '())))
((1 1 E 1) (car/cons x2 (cons x1 '())))
((1 1 E 2 1) (cdr/cons x2 (cons x1 '())))
((1)
 (if-same (equal x2 '?)
   (memb?
     (if (equal x2 '?) (remb (cons x1 '())) (cons x2 (remb (cons x1 '())))))))
((1 A 1)
 (if-nest-A (equal x2 '?) (remb (cons x1 '())) (cons x2 (remb (cons x1 '())))))
((1 E 1)
 (if-nest-E (equal x2 '?) (remb (cons x1 '())) (cons x2 (remb (cons x1 '())))))
((1 A) (memb?/remb1 x1))
((1 E) (memb? (cons x2 (remb (cons x1 '())))))
((1 E Q) (atom/cons x2 (remb (cons x1 '()))))
((1 E)
 (if-false 'nil
   (if (equal (car (cons x2 (remb (cons x1 '())))) '?)
       't
       (memb? (cdr (cons x2 (remb (cons x1 '()))))))))
((1 E Q 1) (car/cons x2 (remb (cons x1 '()))))
((1 E E 1) (cdr/cons x2 (remb (cons x1 '()))))
((1 E) (if-nest-E (equal x2 '?) 't (memb? (remb (cons x1 '())))))
((1 E) (memb?/remb1 x1))
((1) (if-same (equal x2 '?) 'nil))
(() (equal-same 'nil))))))
```

Chapter 6 Proofs

```
(defun dethm.memb?/remb ()
  (J-Bob/define (dethm.memb?/remb2)
    '(((dethm memb?/remb (xs)
         (equal (memb? (remb xs)) 'nil))
       (list-induction xs)
       ((A 1 1) (remb xs))
       ((A 1 1)
        (if-nest-A (atom xs)
          '()
          (if (equal (car xs) '?) (remb (cdr xs)) (cons (car xs) (remb (cdr xs))))))
       ((A 1) (memb? '()))
       ((A 1 Q) (atom '()))
       ((A 1) (if-true 'nil (if (equal (car '()) '?) 't (memb? (cdr '())))))
       ((A) (equal-same 'nil))
       ((E A 1 1) (remb xs))
```

```
((E A 1 1)
 (if-nest-E (atom xs)
   '()
   (if (equal (car xs) '?) (remb (cdr xs)) (cons (car xs) (remb (cdr xs))))))
((E A 1)
 (if-same (equal (car xs) '?)
   (memb?
     (if (equal (car xs) '?) (remb (cdr xs)) (cons (car xs) (remb (cdr xs)))))))
((E A 1 A 1)
 (if-nest-A (equal (car xs) '?) (remb (cdr xs)) (cons (car xs) (remb (cdr xs)))))
((E A 1 E 1)
 (if-nest-E (equal (car xs) '?) (remb (cdr xs)) (cons (car xs) (remb (cdr xs)))))
((E A 1 A) (equal-if (memb? (remb (cdr xs))) 'nil))
((E A 1 E) (memb? (cons (car xs) (remb (cdr xs)))))
((E A 1 E Q) (atom/cons (car xs) (remb (cdr xs))))
((E A 1 E)
 (if-false 'nil
   (if (equal (car (cons (car xs) (remb (cdr xs)))) '?)
       't
       (memb? (cdr (cons (car xs) (remb (cdr xs))))))))
((E A 1 E Q 1) (car/cons (car xs) (remb (cdr xs))))
((E A 1 E E 1) (cdr/cons (car xs) (remb (cdr xs))))
((E A 1 E) (if-nest-E (equal (car xs) '?) 't (memb? (remb (cdr xs)))))
((E A 1 E) (equal-if (memb? (remb (cdr xs))) 'nil))
((E A 1) (if-same (equal (car xs) '?) 'nil))
((E A) (equal-same 'nil))
((E) (if-same (equal (memb? (remb (cdr xs))) 'nil) 't))
(() (if-same (atom xs) 't)))))
```

Chapter 7 Proofs

```
(defun defun.ctx? ()
  (J-Bob/define (dethm.memb?/remb)
    '(((defun ctx? (x)
         (if (atom x) (equal x '?) (if (ctx? (car x)) 't (ctx? (cdr x)))))
       (size x)
       ((Q) (natp/size x))
       (()
        (if-true
          (if (atom x)
              't
              (if (< (size (car x)) (size x))
                  (if (ctx? (car x)) 't (< (size (cdr x)) (size x)))
                  'nil))
          'nil))
       ((E Q) (size/car x))
       ((E A E) (size/cdr x))
       ((E A) (if-same (ctx? (car x)) 't))
       ((E) (if-true 't 'nil))
       (() (if-same (atom x) 't))))))

(defun dethm.ctx?/sub ()
  (J-Bob/define (defun.ctx?)
    '(((dethm ctx?/t (x)
         (if (ctx? x) (equal (ctx? x) 't) 't))
```

```
(star-induction x)
((A A 1) (ctx? x))
((A A 1) (if-nest-A (atom x) (equal x '?) (if (ctx? (car x)) 't (ctx? (cdr x)))))
((A Q) (ctx? x))
((A Q) (if-nest-A (atom x) (equal x '?) (if (ctx? (car x)) 't (ctx? (cdr x)))))
((A A 1 1) (equal-if x '?))
((A A 1) (equal-same '?))
((A A) (equal-same 't))
((A) (if-same (equal x '?) 't))
((E A A A 1) (ctx? x))
((E A A A 1)
 (if-nest-E (atom x) (equal x '?) (if (ctx? (car x)) 't (ctx? (cdr x)))))
((E)
 (if-same (ctx? (car x))
   (if (if (ctx? (car x)) (equal (ctx? (car x)) 't) 't)
       (if (if (ctx? (cdr x)) (equal (ctx? (cdr x)) 't) 't)
           (if (ctx? x) (equal (if (ctx? (car x)) 't (ctx? (cdr x))) 't) 't)
           't)
       't)))
((E A Q) (if-nest-A (ctx? (car x)) (equal (ctx? (car x)) 't) 't))
((E A A A A 1) (if-nest-A (ctx? (car x)) 't (ctx? (cdr x))))
((E E Q) (if-nest-E (ctx? (car x)) (equal (ctx? (car x)) 't) 't))
((E E A A A 1) (if-nest-E (ctx? (car x)) 't (ctx? (cdr x))))
((E A A A A) (equal-same 't))
((E E)
 (if-true
   (if (if (ctx? (cdr x)) (equal (ctx? (cdr x)) 't) 't)
       (if (ctx? x) (equal (ctx? (cdr x)) 't) 't)
       't)
   't))
((E A A A) (if-same (ctx? x) 't))
((E A A) (if-same (if (ctx? (cdr x)) (equal (ctx? (cdr x)) 't) 't) 't))
((E A) (if-same (equal (ctx? (car x)) 't) 't))
((E E A Q) (ctx? x))
((E E A Q)
 (if-nest-E (atom x) (equal x '?) (if (ctx? (car x)) 't (ctx? (cdr x)))))
((E E A Q) (if-nest-E (ctx? (car x)) 't (ctx? (cdr x))))
((E E)
 (if-same (ctx? (cdr x))
   (if (if (ctx? (cdr x)) (equal (ctx? (cdr x)) 't) 't)
       (if (ctx? (cdr x)) (equal (ctx? (cdr x)) 't) 't)
       't)))
((E E A Q) (if-nest-A (ctx? (cdr x)) (equal (ctx? (cdr x)) 't) 't))
((E E A A) (if-nest-A (ctx? (cdr x)) (equal (ctx? (cdr x)) 't) 't))
((E E E Q) (if-nest-E (ctx? (cdr x)) (equal (ctx? (cdr x)) 't) 't))
((E E E A) (if-nest-E (ctx? (cdr x)) (equal (ctx? (cdr x)) 't) 't))
((E E E) (if-same 't 't))
((E E A A 1) (equal-if (ctx? (cdr x)) 't))
((E E A A) (equal-same 't))
((E E A) (if-same (equal (ctx? (cdr x)) 't) 't))
((E E) (if-same (ctx? (cdr x)) 't))
((E) (if-same (ctx? (car x)) 't))
(() (if-same (atom x) 't)))
((dethm ctx?/sub (x y)
  (if (ctx? x) (if (ctx? y) (equal (ctx? (sub x y)) 't) 't) 't))
```

```
(star-induction y)
(()
 (if-same (ctx? x)
   (if (atom y)
       (if (ctx? x) (if (ctx? y) (equal (ctx? (sub x y)) 't) 't) 't)
       (if (if (ctx? x)
               (if (ctx? (car y)) (equal (ctx? (sub x (car y))) 't) 't)
               't)
           (if (if (ctx? x)
                   (if (ctx? (cdr y)) (equal (ctx? (sub x (cdr y))) 't) 't)
                   't)
               (if (ctx? x) (if (ctx? y) (equal (ctx? (sub x y)) 't) 't) 't)
               't)
           't))))
((A A) (if-nest-A (ctx? x) (if (ctx? y) (equal (ctx? (sub x y)) 't) 't) 't))
((A E Q)
 (if-nest-A (ctx? x) (if (ctx? (car y)) (equal (ctx? (sub x (car y))) 't) 't) 't))
((A E A Q)
 (if-nest-A (ctx? x) (if (ctx? (cdr y)) (equal (ctx? (sub x (cdr y))) 't) 't) 't))
((A E A A) (if-nest-A (ctx? x) (if (ctx? y) (equal (ctx? (sub x y)) 't) 't) 't))
((E A) (if-nest-E (ctx? x) (if (ctx? y) (equal (ctx? (sub x y)) 't) 't) 't))
((E E Q)
 (if-nest-E (ctx? x) (if (ctx? (car y)) (equal (ctx? (sub x (car y))) 't) 't) 't))
((E E A Q)
 (if-nest-E (ctx? x) (if (ctx? (cdr y)) (equal (ctx? (sub x (cdr y))) 't) 't) 't))
((E E A A) (if-nest-E (ctx? x) (if (ctx? y) (equal (ctx? (sub x y)) 't) 't) 't))
((E E A) (if-same 't 't))
((E E) (if-same 't 't))
((E) (if-same (atom y) 't))
((A A A 1 1) (sub x y))
((A A A 1 1)
 (if-nest-A (atom y)
   (if (equal y '?) x y)
   (cons (sub x (car y)) (sub x (cdr y)))))
((A A A) (if-same (equal y '?) (equal (ctx? (if (equal y '?) x y)) 't)))
((A A A A 1 1) (if-nest-A (equal y '?) x y))
((A A A E 1 1) (if-nest-E (equal y '?) x y))
((A A A A 1) (ctx?/t x))
((A A A A) (equal-same 't))
((A A A E 1) (ctx?/t y))
((A A A E) (equal-same 't))
((A A A) (if-same (equal y '?) 't))
((A A) (if-same (ctx? y) 't))
((A E A A A 1 1) (sub x y))
((A E A A A 1 1)
 (if-nest-E (atom y)
   (if (equal y '?) x y)
   (cons (sub x (car y)) (sub x (cdr y)))))
((A E A A A 1) (ctx? (cons (sub x (car y)) (sub x (cdr y)))))
((A E A A A 1 Q) (atom/cons (sub x (car y)) (sub x (cdr y))))
((A E A A A 1 E Q 1) (car/cons (sub x (car y)) (sub x (cdr y))))
((A E A A A 1 E E 1) (cdr/cons (sub x (car y)) (sub x (cdr y))))
```

```
((A E A A A 1)
 (if-false (equal (cons (sub x (car y)) (sub x (cdr y))) '?)
   (if (ctx? (sub x (car y))) 't (ctx? (sub x (cdr y))))))
((A E A A Q) (ctx? y))
((A E A A Q)
 (if-nest-E (atom y) (equal y '?) (if (ctx? (car y)) 't (ctx? (cdr y)))))
((A E)
 (if-same (ctx? (car y))
   (if (if (ctx? (car y)) (equal (ctx? (sub x (car y))) 't) 't)
       (if (if (ctx? (cdr y)) (equal (ctx? (sub x (cdr y))) 't) 't)
           (if (if (ctx? (car y)) 't (ctx? (cdr y)))
               (equal (if (ctx? (sub x (car y))) 't (ctx? (sub x (cdr y)))) 't)
               't)
           't)
       't)))
((A E A Q) (if-nest-A (ctx? (car y)) (equal (ctx? (sub x (car y))) 't) 't))
((A E A A A Q) (if-nest-A (ctx? (car y)) 't (ctx? (cdr y))))
((A E E Q) (if-nest-E (ctx? (car y)) (equal (ctx? (sub x (car y))) 't) 't))
((A E E A A Q) (if-nest-E (ctx? (car y)) 't (ctx? (cdr y))))
((A E A A A)
 (if-true (equal (if (ctx? (sub x (car y))) 't (ctx? (sub x (cdr y)))) 't) 't))
((A E E)
 (if-true
   (if (if (ctx? (cdr y)) (equal (ctx? (sub x (cdr y))) 't) 't)
       (if (ctx? (cdr y))
           (equal (if (ctx? (sub x (car y))) 't (ctx? (sub x (cdr y)))) 't)
           't)
       't)
   't))
((A E A A A 1 Q) (equal-if (ctx? (sub x (car y))) 't))
((A E A A A 1) (if-true 't (ctx? (sub x (cdr y)))))
((A E A A A) (equal-same 't))
((A E A A) (if-same (if (ctx? (cdr y)) (equal (ctx? (sub x (cdr y))) 't) 't) 't))
((A E A) (if-same (equal (ctx? (sub x (car y))) 't) 't))
((A E E)
 (if-same (ctx? (cdr y))
   (if (if (ctx? (cdr y)) (equal (ctx? (sub x (cdr y))) 't) 't)
       (if (ctx? (cdr y))
           (equal (if (ctx? (sub x (car y))) 't (ctx? (sub x (cdr y)))) 't)
           't)
       't)))
((A E E A Q) (if-nest-A (ctx? (cdr y)) (equal (ctx? (sub x (cdr y))) 't) 't))
((A E E A A)
 (if-nest-A (ctx? (cdr y))
   (equal (if (ctx? (sub x (car y))) 't (ctx? (sub x (cdr y)))) 't)
   't))
((A E E E Q) (if-nest-E (ctx? (cdr y)) (equal (ctx? (sub x (cdr y))) 't) 't))
((A E E E A)
 (if-nest-E (ctx? (cdr y))
   (equal (if (ctx? (sub x (car y))) 't (ctx? (sub x (cdr y)))) 't)
   't))
((A E E E) (if-same 't 't))
((A E E A A 1 E) (equal-if (ctx? (sub x (cdr y))) 't))
((A E E A A 1) (if-same (ctx? (sub x (car y))) 't))
```

```
      ((A E E A A) (equal-same 't))
      ((A E E A) (if-same (equal (ctx? (sub x (cdr y))) 't) 't))
      ((A E E) (if-same (ctx? (cdr y)) 't))
      ((A E) (if-same (ctx? (car y)) 't))
      ((A) (if-same (atom y) 't))
      (() (if-same (ctx? x) 't))))))
```

Chapter 8 Proofs

```
(defun defun.member? ()
  (J-Bob/define (dethm.ctx?/sub)
    '(((defun member? (x ys)
         (if (atom ys) 'nil (if (equal x (car ys)) 't (member? x (cdr ys)))))
       (size ys)
       ((Q) (natp/size ys))
       (()
        (if-true
          (if (atom ys) 't (if (equal x (car ys)) 't (< (size (cdr ys)) (size ys))))
          'nil))
       ((E E) (size/cdr ys))
       ((E) (if-same (equal x (car ys)) 't))
       (() (if-same (atom ys) 't))))))
(defun defun.set? ()
  (J-Bob/define (defun.member?)
    '(((defun set? (xs)
         (if (atom xs) 't (if (member? (car xs) (cdr xs)) 'nil (set? (cdr xs)))))
       (size xs)
       ((Q) (natp/size xs))
       (()
        (if-true
          (if (atom xs)
              't
              (if (member? (car xs) (cdr xs)) 't (< (size (cdr xs)) (size xs))))
          'nil))
       ((E E) (size/cdr xs))
       ((E) (if-same (member? (car xs) (cdr xs)) 't))
       (() (if-same (atom xs) 't))))))
(defun defun.add-atoms ()
  (J-Bob/define (defun.set?)
    '(((defun add-atoms (x ys)
         (if (atom x)
             (if (member? x ys) ys (cons x ys))
             (add-atoms (car x) (add-atoms (cdr x) ys))))
       (size x)
       ((Q) (natp/size x))
       (()
        (if-true
          (if (atom x)
              't
              (if (< (size (car x)) (size x)) (< (size (cdr x)) (size x)) 'nil))
          'nil))
       ((E Q) (size/car x))
       ((E A) (size/cdr x))
       ((E) (if-true 't 'nil))
       (() (if-same (atom x) 't))))))
```

```
(defun defun.atoms ()
  (J-Bob/define (defun.add-atoms)
    '(((defun atoms (x)
         (add-atoms x '()))
       nil))))
```

Chapter 9 Proofs

```
(defun dethm.set?/atoms.attempt ()
  (J-Bob/prove (defun.atoms)
    '(((dethm set?/add-atoms (a)
         (equal (set? (add-atoms a '())) 't))
       (star-induction a)
       ((E A A 1 1) (add-atoms a '())))
      ((dethm set?/atoms (a)
         (equal (set? (atoms a)) 't))
       nil
       ((1 1) (atoms a))
       ((1) (set?/add-atoms a))
       (() (equal-same 't))))))
```

```
(defun dethm.set?/atoms ()
  (J-Bob/define (defun.atoms)
    '(((dethm set?/t (xs)
         (if (set? xs) (equal (set? xs) 't) 't))
       (list-induction xs)
       ((A A 1) (set? xs))
       ((A A 1)
        (if-nest-A (atom xs) 't (if (member? (car xs) (cdr xs)) 'nil (set? (cdr xs)))))
       ((A A) (equal-same 't))
       ((A) (if-same (set? xs) 't))
       ((E A A 1) (set? xs))
       ((E A A 1)
        (if-nest-E (atom xs) 't (if (member? (car xs) (cdr xs)) 'nil (set? (cdr xs)))))
       ((E A Q) (set? xs))
       ((E A Q)
        (if-nest-E (atom xs) 't (if (member? (car xs) (cdr xs)) 'nil (set? (cdr xs)))))
       ((E A)
        (if-same (member? (car xs) (cdr xs))
          (if (if (member? (car xs) (cdr xs)) 'nil (set? (cdr xs)))
              (equal (if (member? (car xs) (cdr xs)) 'nil (set? (cdr xs))) 't)
              't)))
       ((E A A Q) (if-nest-A (member? (car xs) (cdr xs)) 'nil (set? (cdr xs))))
       ((E A A A 1) (if-nest-A (member? (car xs) (cdr xs)) 'nil (set? (cdr xs))))
       ((E A E Q) (if-nest-E (member? (car xs) (cdr xs)) 'nil (set? (cdr xs))))
       ((E A E A 1) (if-nest-E (member? (car xs) (cdr xs)) 'nil (set? (cdr xs))))
       ((E A A) (if-false (equal 'nil 't) 't))
       ((E)
        (if-same (set? (cdr xs))
          (if (if (set? (cdr xs)) (equal (set? (cdr xs)) 't) 't)
              (if (member? (car xs) (cdr xs))
                  't
                  (if (set? (cdr xs)) (equal (set? (cdr xs)) 't) 't))
              't)))
       ((E A Q) (if-nest-A (set? (cdr xs)) (equal (set? (cdr xs)) 't) 't))
       ((E A A E) (if-nest-A (set? (cdr xs)) (equal (set? (cdr xs)) 't) 't))
```

```
 ((E E Q) (if-nest-E (set? (cdr xs)) (equal (set? (cdr xs)) 't) 't))
 ((E E A E) (if-nest-E (set? (cdr xs)) (equal (set? (cdr xs)) 't) 't))
 ((E E A) (if-same (member? (car xs) (cdr xs)) 't))
 ((E E) (if-same 't 't))
 ((E A A E 1) (equal-if (set? (cdr xs)) 't))
 ((E A A E) (equal-same 't))
 ((E A A) (if-same (member? (car xs) (cdr xs)) 't))
 ((E A) (if-same (equal (set? (cdr xs)) 't) 't))
 ((E) (if-same (set? (cdr xs)) 't))
 (() (if-same (atom xs) 't)))
((dethm set?/nil (xs)
   (if (set? xs) 't (equal (set? xs) 'nil)))
 (list-induction xs)
 ((A Q) (set? xs))
 ((A Q)
  (if-nest-A (atom xs) 't (if (member? (car xs) (cdr xs)) 'nil (set? (cdr xs)))))
 ((A) (if-true 't (equal (set? xs) 'nil)))
 ((E A E 1) (set? xs))
 ((E A E 1)
  (if-nest-E (atom xs) 't (if (member? (car xs) (cdr xs)) 'nil (set? (cdr xs)))))
 ((E A Q) (set? xs))
 ((E A Q)
  (if-nest-E (atom xs) 't (if (member? (car xs) (cdr xs)) 'nil (set? (cdr xs)))))
 ((E A)
  (if-same (member? (car xs) (cdr xs))
    (if (if (member? (car xs) (cdr xs)) 'nil (set? (cdr xs)))
        't
        (equal (if (member? (car xs) (cdr xs)) 'nil (set? (cdr xs))) 'nil))))
 ((E A A Q) (if-nest-A (member? (car xs) (cdr xs)) 'nil (set? (cdr xs))))
 ((E A A E 1) (if-nest-A (member? (car xs) (cdr xs)) 'nil (set? (cdr xs))))
 ((E A E Q) (if-nest-E (member? (car xs) (cdr xs)) 'nil (set? (cdr xs))))
 ((E A E E 1) (if-nest-E (member? (car xs) (cdr xs)) 'nil (set? (cdr xs))))
 ((E A A E) (equal-same 'nil))
 ((E A A) (if-same 'nil 't))
 ((E)
  (if-same (set? (cdr xs))
    (if (if (set? (cdr xs)) 't (equal (set? (cdr xs)) 'nil))
        (if (member? (car xs) (cdr xs))
            't
            (if (set? (cdr xs)) 't (equal (set? (cdr xs)) 'nil)))
        't)))
 ((E A Q) (if-nest-A (set? (cdr xs)) 't (equal (set? (cdr xs)) 'nil)))
 ((E A A E) (if-nest-A (set? (cdr xs)) 't (equal (set? (cdr xs)) 'nil)))
 ((E E Q) (if-nest-E (set? (cdr xs)) 't (equal (set? (cdr xs)) 'nil)))
 ((E E A E) (if-nest-E (set? (cdr xs)) 't (equal (set? (cdr xs)) 'nil)))
 ((E A A) (if-same (member? (car xs) (cdr xs)) 't))
 ((E A) (if-same 't 't))
 ((E E A E 1) (equal-if (set? (cdr xs)) 'nil))
 ((E E A E) (equal-same 'nil))
 ((E E A) (if-same (member? (car xs) (cdr xs)) 't))
 ((E E) (if-same (equal (set? (cdr xs)) 'nil) 't))
 ((E) (if-same (set? (cdr xs)) 't))
 (() (if-same (atom xs) 't)))
((dethm set?/add-atoms (a bs)
   (if (set? bs) (equal (set? (add-atoms a bs)) 't) 't))
```

```
(add-atoms a bs)
((A A 1 1) (add-atoms a bs))
((A A 1 1)
 (if-nest-A (atom a)
   (if (member? a bs) bs (cons a bs))
   (add-atoms (car a) (add-atoms (cdr a) bs))))
((A A 1) (if-same (member? a bs) (set? (if (member? a bs) bs (cons a bs)))))
((A A 1 A 1) (if-nest-A (member? a bs) bs (cons a bs)))
((A A 1 E 1) (if-nest-E (member? a bs) bs (cons a bs)))
((A A 1 A) (set?/t bs))
((A A 1 E) (set? (cons a bs)))
((A A 1 E Q) (atom/cons a bs))
((A A 1 E E Q 1) (car/cons a bs))
((A A 1 E E Q 2) (cdr/cons a bs))
((A A 1 E E E 1) (cdr/cons a bs))
((A A 1 E) (if-false 't (if (member? a bs) 'nil (set? bs))))
((A A 1 E) (if-nest-E (member? a bs) 'nil (set? bs)))
((A A 1 E) (set?/t bs))
((A A 1) (if-same (member? a bs) 't))
((A A) (equal-same 't))
((A) (if-same (set? bs) 't))
((E)
 (if-same (set? bs)
   (if (if (set? (add-atoms (cdr a) bs))
           (equal (set? (add-atoms (car a) (add-atoms (cdr a) bs))) 't)
           't)
       (if (if (set? bs) (equal (set? (add-atoms (cdr a) bs)) 't) 't)
           (if (set? bs) (equal (set? (add-atoms a bs)) 't) 't)
           't)
       't)))
((E A A Q) (if-nest-A (set? bs) (equal (set? (add-atoms (cdr a) bs)) 't) 't))
((E A A A) (if-nest-A (set? bs) (equal (set? (add-atoms a bs)) 't) 't))
((E E A Q) (if-nest-E (set? bs) (equal (set? (add-atoms (cdr a) bs)) 't) 't))
((E E A A) (if-nest-E (set? bs) (equal (set? (add-atoms a bs)) 't) 't))
((E E A) (if-same 't 't))
((E E)
 (if-same
   (if (set? (add-atoms (cdr a) bs))
       (equal (set? (add-atoms (car a) (add-atoms (cdr a) bs))) 't)
       't)
   't))
((E A)
 (if-same (set? (add-atoms (cdr a) bs))
   (if (if (set? (add-atoms (cdr a) bs))
           (equal (set? (add-atoms (car a) (add-atoms (cdr a) bs))) 't)
           't)
       (if (equal (set? (add-atoms (cdr a) bs)) 't)
           (equal (set? (add-atoms a bs)) 't)
           't)
       't)))
((E A A Q)
 (if-nest-A (set? (add-atoms (cdr a) bs))
   (equal (set? (add-atoms (car a) (add-atoms (cdr a) bs))) 't)
   't))
```

```
((E A E Q)
 (if-nest-E (set? (add-atoms (cdr a) bs))
   (equal (set? (add-atoms (car a) (add-atoms (cdr a) bs))) 't)
   't))
((E A E)
 (if-true
   (if (equal (set? (add-atoms (cdr a) bs)) 't)
       (equal (set? (add-atoms a bs)) 't)
       't)
   't))
((E A A A Q 1) (set?/t (add-atoms (cdr a) bs)))
((E A E Q 1) (set?/nil (add-atoms (cdr a) bs)))
((E A A A Q) (equal 't 't))
((E A E Q) (equal 'nil 't))
((E A A A) (if-true (equal (set? (add-atoms a bs)) 't) 't))
((E A E) (if-false (equal (set? (add-atoms a bs)) 't) 't))
((E A A A 1 1) (add-atoms a bs))
((E A A A 1 1)
 (if-nest-E (atom a)
   (if (member? a bs) bs (cons a bs))
   (add-atoms (car a) (add-atoms (cdr a) bs))))
((E A A A 1) (equal-if (set? (add-atoms (car a) (add-atoms (cdr a) bs))) 't))
((E A A A) (equal-same 't))
((E A A)
 (if-same (equal (set? (add-atoms (car a) (add-atoms (cdr a) bs))) 't) 't))
((E A) (if-same (set? (add-atoms (cdr a) bs)) 't))
((E) (if-same (set? bs) 't))
(() (if-same (atom a) 't)))
((dethm set?/atoms (a)
  (equal (set? (atoms a)) 't))
nil
((1 1) (atoms a))
(() (if-true (equal (set? (add-atoms a '())) 't) 't))
((Q) (if-true 't (if (member? (car '()) (cdr '())) 'nil (set? (cdr '())))))
((Q Q) (atom '()))
((Q) (set? '()))
((A 1) (set?/add-atoms a '()))
((A) (equal-same 't))
(() (if-same (set? '()) 't)))))
```

Chapter 10 Proofs

```
(defun defun.rotate ()
  (J-Bob/define (dethm.set?/atoms)
    '(((defun rotate (x)
         (cons (car (car x)) (cons (cdr (car x)) (cdr x))))
       nil))))

(defun dethm.rotate/cons ()
  (J-Bob/define (defun.rotate)
    '(((dethm rotate/cons (x y z)
         (equal (rotate (cons (cons x y) z)) (cons x (cons y z))))
       nil
       ((1) (rotate (cons (cons x y) z)))
       ((1 1 1) (car/cons (cons x y) z))
```

```
      ((1 1) (car/cons x y))
      ((1 2 1 1) (car/cons (cons x y) z))
      ((1 2 1) (cdr/cons x y))
      ((1 2 2) (cdr/cons (cons x y) z))
      (() (equal-same (cons x (cons y z))))))))

(defun defun.align.attempt ()
  (J-Bob/prove (dethm.rotate/cons)
    '(((defun align (x)
         (if (atom x)
             x
             (if (atom (car x)) (cons (car x) (align (cdr x))) (align (rotate x)))))
       (size x)
       ((Q) (natp/size x))
       (()
        (if-true
          (if (atom x)
              't
              (if (atom (car x))
                  (< (size (cdr x)) (size x))
                  (< (size (rotate x)) (size x))))
          'nil))
       ((E A) (size/cdr x))
       ((E E 1 1 1) (cons/car+cdr x))
       ((E E 2 1) (cons/car+cdr x))
       ((E E 1 1 1 1) (cons/car+cdr (car x)))
       ((E E 2 1 1) (cons/car+cdr (car x)))
       ((E E 1 1) (rotate/cons (car (car x)) (cdr (car x)) (cdr x)))))))

(defun defun.wt ()
  (J-Bob/define (dethm.rotate/cons)
    '(((defun wt (x)
         (if (atom x) '1 (+ (+ (wt (car x)) (wt (car x))) (wt (cdr x)))))
       (size x)
       ((Q) (natp/size x))
       (()
        (if-true
          (if (atom x)
              't
              (if (< (size (car x)) (size x)) (< (size (cdr x)) (size x)) 'nil))
          'nil))
       ((E Q) (size/car x))
       ((E A) (size/cdr x))
       ((E) (if-true 't 'nil))
       (() (if-same (atom x) 't))))))

(defun defun.align ()
  (J-Bob/define (defun.wt)
    '(((dethm natp/wt (x)
         (equal (natp (wt x)) 't))
       (star-induction x)
       ((A 1 1) (wt x))
       ((A 1 1) (if-nest-A (atom x) '1 (+ (+ (wt (car x)) (wt (car x))) (wt (cdr x)))))
       ((A 1) (natp '1))
       ((A) (equal-same 't))
       ((E A A 1 1) (wt x))
```

```
((E A A 1 1)
 (if-nest-E (atom x) '1 (+ (+ (wt (car x)) (wt (car x))) (wt (cdr x)))))
((E A A)
 (if-true (equal (natp (+ (+ (wt (car x)) (wt (car x))) (wt (cdr x)))) 't) 't))
((E A A Q) (equal-if (natp (wt (car x))) 't))
((E A A A)
 (if-true (equal (natp (+ (+ (wt (car x)) (wt (car x))) (wt (cdr x)))) 't) 't))
((E A A A Q) (natp/+ (wt (car x)) (wt (car x))))
((E A A Q) (equal-if (natp (wt (car x))) 't))
((E A A Q) (equal-if (natp (wt (cdr x))) 't))
((E A A A A 1) (natp/+ (+ (wt (car x)) (wt (car x))) (wt (cdr x))))
((E A A A A) (equal-same 't))
((E A A A) (if-same (natp (+ (wt (car x)) (wt (car x)))) 't))
((E A A) (if-same (natp (wt (cdr x))) 't))
((E A) (if-same (equal (natp (wt (cdr x))) 't) 't))
((E) (if-same (equal (natp (wt (car x))) 't) 't))
(() (if-same (atom x) 't)))
((dethm positive/wt (x)
  (equal (< '0 (wt x)) 't))
(star-induction x)
((A 1 2) (wt x))
((A 1 2) (if-nest-A (atom x) '1 (+ (+ (wt (car x)) (wt (car x))) (wt (cdr x)))))
((A 1) (< '0 '1))
((A) (equal-same 't))
((E A A 1 2) (wt x))
((E A A 1 2)
 (if-nest-E (atom x) '1 (+ (+ (wt (car x)) (wt (car x))) (wt (cdr x)))))
((E A A)
 (if-true (equal (< '0 (+ (+ (wt (car x)) (wt (car x))) (wt (cdr x)))) 't) 't))
((E A A Q) (equal-if (< '0 (wt (car x))) 't))
((E A A A)
 (if-true (equal (< '0 (+ (+ (wt (car x)) (wt (car x))) (wt (cdr x)))) 't) 't))
((E A A A Q) (positives-+ (wt (car x)) (wt (car x))))
((E A A Q) (equal-if (< '0 (wt (car x))) 't))
((E A A Q) (equal-if (< '0 (wt (cdr x))) 't))
((E A A A A 1) (positives-+ (+ (wt (car x)) (wt (car x))) (wt (cdr x))))
((E A A A A) (equal-same 't))
((E A A A) (if-same (< '0 (+ (wt (car x)) (wt (car x)))) 't))
((E A A) (if-same (< '0 (wt (cdr x))) 't))
((E A) (if-same (equal (< '0 (wt (cdr x))) 't) 't))
((E) (if-same (equal (< '0 (wt (car x))) 't) 't))
(() (if-same (atom x) 't)))
((defun align (x)
  (if (atom x)
      x
      (if (atom (car x)) (cons (car x) (align (cdr x))) (align (rotate x)))))
(wt x)
((Q) (natp/wt x))
(()
 (if-true
   (if (atom x)
       't
       (if (atom (car x)) (< (wt (cdr x)) (wt x)) (< (wt (rotate x)) (wt x))))
   'nil))
((E A 2) (wt x))
```

```
((E A 2) (if-nest-E (atom x) '1 (+ (+ (wt (car x)) (wt (car x))) (wt (cdr x)))))
((E A)
 (if-true (< (wt (cdr x)) (+ (+ (wt (car x)) (wt (car x))) (wt (cdr x)))) 't))
((E A Q) (natp/wt (cdr x)))
((E A A 1) (identity-+ (wt (cdr x))))
((E A A) (common-addends-< '0 (+ (wt (car x)) (wt (car x))) (wt (cdr x))))
((E A Q) (natp/wt (cdr x)))
((E A Q) (positive/wt (car x)))
((E A A) (positives-+ (wt (car x)) (wt (car x))))
((E A) (if-same (< '0 (wt (car x))) 't))
((E E 1 1) (rotate x))
((E E 1) (wt (cons (car (car x)) (cons (cdr (car x)) (cdr x)))))
((E E 1 Q) (atom/cons (car (car x)) (cons (cdr (car x)) (cdr x))))
((E E 1)
 (if-false '1
   (+ (+ (wt (car (cons (car (car x)) (cons (cdr (car x)) (cdr x)))))
         (wt (car (cons (car (car x)) (cons (cdr (car x)) (cdr x))))))
      (wt (cdr (cons (car (car x)) (cons (cdr (car x)) (cdr x))))))))
((E E 1 1 1 1) (car/cons (car (car x)) (cons (cdr (car x)) (cdr x))))
((E E 1 1 2 1) (car/cons (car (car x)) (cons (cdr (car x)) (cdr x))))
((E E 1 2 1) (cdr/cons (car (car x)) (cons (cdr (car x)) (cdr x))))
((E E 1 2) (wt (cons (cdr (car x)) (cdr x))))
((E E 1 2 Q) (atom/cons (cdr (car x)) (cdr x)))
((E E 1 2)
 (if-false '1
   (+ (+ (wt (car (cons (cdr (car x)) (cdr x))))
         (wt (car (cons (cdr (car x)) (cdr x)))))
      (wt (cdr (cons (cdr (car x)) (cdr x)))))))
((E E 1 2 1 1 1) (car/cons (cdr (car x)) (cdr x)))
((E E 1 2 1 2 1) (car/cons (cdr (car x)) (cdr x)))
((E E 1 2 2 1) (cdr/cons (cdr (car x)) (cdr x)))
((E E 2) (wt x))
((E E 2) (if-nest-E (atom x) '1 (+ (+ (wt (car x)) (wt (car x))) (wt (cdr x)))))
((E E 2 1 1) (wt (car x)))
((E E 2 1 1)
 (if-nest-E (atom (car x))
   '1
   (+ (+ (wt (car (car x))) (wt (car (car x)))) (wt (cdr (car x))))))
((E E 2 1 2) (wt (car x)))
((E E 2 1 2)
 (if-nest-E (atom (car x))
   '1
   (+ (+ (wt (car (car x))) (wt (car (car x)))) (wt (cdr (car x))))))
((E E 1)
 (associate-+
   (+ (wt (car (car x))) (wt (car (car x))))
   (+ (wt (cdr (car x))) (wt (cdr (car x))))
   (wt (cdr x))))
((E E)
 (common-addends-<
   (+ (+ (wt (car (car x))) (wt (car (car x))))
      (+ (wt (cdr (car x))) (wt (cdr (car x)))))
   (+ (+ (+ (wt (car (car x))) (wt (car (car x)))) (wt (cdr (car x))))
      (+ (+ (wt (car (car x))) (wt (car (car x)))) (wt (cdr (car x)))))
   (wt (cdr x))))
```

```
        ((E E 1)
         (associate-+
           (+ (wt (car (car x))) (wt (car (car x))))
           (wt (cdr (car x)))
           (wt (cdr (car x)))))
        ((E E 1)
         (commute-+
           (+ (+ (wt (car (car x))) (wt (car (car x)))) (wt (cdr (car x))))
           (wt (cdr (car x)))))
        ((E E)
         (common-addends-<
           (wt (cdr (car x)))
           (+ (+ (wt (car (car x))) (wt (car (car x)))) (wt (cdr (car x))))
           (+ (+ (wt (car (car x))) (wt (car (car x)))) (wt (cdr (car x))))))
        ((E E)
         (if-true
           (< (wt (cdr (car x)))
              (+ (+ (wt (car (car x))) (wt (car (car x)))) (wt (cdr (car x)))))
           't))
        ((E E Q) (natp/wt (cdr (car x))))
        ((E E A 1) (identity-+ (wt (cdr (car x)))))
        ((E E A)
         (common-addends-<
           '0
           (+ (wt (car (car x))) (wt (car (car x))))
           (wt (cdr (car x)))))
        ((E E Q) (natp/wt (cdr (car x))))
        ((E E Q) (positive/wt (car (car x))))
        ((E E A) (positives-+ (wt (car (car x))) (wt (car (car x)))))
        ((E E) (if-same (< '0 (wt (car (car x)))) 't))
        ((E) (if-same (atom (car x)) 't))
        (() (if-same (atom x) 't))))))

(defun dethm.align/align ()
  (J-Bob/define (defun.align)
    '(((dethm align/align (x)
         (equal (align (align x)) (align x)))
        (align x)
        ((A 1 1) (align x))
        ((A 1 1)
         (if-nest-A (atom x)
           x
           (if (atom (car x)) (cons (car x) (align (cdr x))) (align (rotate x)))))
        ((A 2) (align x))
        ((A 2)
         (if-nest-A (atom x)
           x
           (if (atom (car x)) (cons (car x) (align (cdr x))) (align (rotate x)))))
        ((A 1) (align x))
        ((A 1)
         (if-nest-A (atom x)
           x
           (if (atom (car x)) (cons (car x) (align (cdr x))) (align (rotate x)))))
        ((A) (equal-same x))
        ((E A A 1 1) (align x))
```

```
((E A A 1 1)
 (if-nest-E (atom x)
   x
   (if (atom (car x)) (cons (car x) (align (cdr x))) (align (rotate x)))))
((E A A 1 1)
 (if-nest-A (atom (car x)) (cons (car x) (align (cdr x))) (align (rotate x))))
((E A A 2) (align x))
((E A A 2)
 (if-nest-E (atom x)
   x
   (if (atom (car x)) (cons (car x) (align (cdr x))) (align (rotate x)))))
((E A A 2)
 (if-nest-A (atom (car x)) (cons (car x) (align (cdr x))) (align (rotate x))))
((E A A 1) (align (cons (car x) (align (cdr x)))))
((E A A 1 Q) (atom/cons (car x) (align (cdr x))))
((E A A 1 E Q 1) (car/cons (car x) (align (cdr x))))
((E A A 1 E A 1) (car/cons (car x) (align (cdr x))))
((E A A 1 E A 2 1) (cdr/cons (car x) (align (cdr x))))
((E A A 1)
 (if-false (cons (car x) (align (cdr x)))
   (if (atom (car x))
       (cons (car x) (align (align (cdr x))))
       (align (rotate (cons (car x) (align (cdr x))))))))
((E A A 1)
 (if-nest-A (atom (car x))
   (cons (car x) (align (align (cdr x))))
   (align (rotate (cons (car x) (align (cdr x)))))))
((E A A 1 2) (equal-if (align (align (cdr x))) (align (cdr x))))
((E A A) (equal-same (cons (car x) (align (cdr x)))))
((E A) (if-same (equal (align (align (cdr x))) (align (cdr x))) 't))
((E E A 1 1) (align x))
((E E A 1 1)
 (if-nest-E (atom x)
   x
   (if (atom (car x)) (cons (car x) (align (cdr x))) (align (rotate x)))))
((E E A 1 1)
 (if-nest-E (atom (car x)) (cons (car x) (align (cdr x))) (align (rotate x))))
((E E A 2) (align x))
((E E A 2)
 (if-nest-E (atom x)
   x
   (if (atom (car x)) (cons (car x) (align (cdr x))) (align (rotate x)))))
((E E A 2)
 (if-nest-E (atom (car x)) (cons (car x) (align (cdr x))) (align (rotate x))))
((E E A 1) (equal-if (align (align (rotate x))) (align (rotate x))))
((E E A) (equal-same (align (rotate x))))
((E E) (if-same (equal (align (align (rotate x))) (align (rotate x))) 't))
((E) (if-same (atom (car x)) 't))
(() (if-same (atom x) 't)))))
```

The Little Assistant

The definition of J-Bob follows; it is also at `http://the-little-prover.org/`. J-Bob is defined in its own language. We first present an implementation of this language in both ACL2 and Scheme; we encourage the reader to implement J-Bob and its language in any other language of choice. There will necessarily be differences between implementations in different languages. For example, in ACL2, (equal 'nil '()) is equal to 't while in Scheme it is equal to 'nil. We have chosen the examples in this book so that the difference never arises. Both implementations of J-Bob are correct and self-consistent because they implement the nine operators to be consistent with J-Bob's axioms.

ACL2. We have chosen the programs in this book to be mostly compatible with the ACL2 theorem prover, on which much of this book is based. Expressions, function definitions, and eight of our nine built-in operators are compatible with ACL2. Both dethm and size must be defined in ACL2:

```
(defun if->implies (exp hyps)
  (case-match exp
    (('if Q A E)
     (append
       (if->implies A `(,@hyps ,Q))
       (if->implies E `(,@hyps (not ,Q)))))
    (('equal X Y)
     `((:rewrite :corollary
         (implies (and ,@hyps)
           (equal ,X ,Y)))))
    (& '())))
```

```
(defmacro dethm (name args body)
  (declare (ignore args))
  (let ((rules (if->implies body '())))
    `(defthm ,name ,body
       :rule-classes ,rules)))

(defun size (x)
  (if (atom x)
    '0
    (+ '1 (size (car x)) (size (cdr x)))))
```

Scheme. The language of programs in this book can be defined in Scheme by redefining if to operate on 't and 'nil rather than #t and #f, defining the missing built-in operators, and by changing some existing operators to be *total*: to return a value no matter what input they are given. Here is the code to do so:

```
(define s.car car)
(define s.cdr cdr)
(define s.+ +)
(define s.< <)
(define (num x) (if (number? x) x 0))

(define (if/nil Q A E)
  (if (equal? Q 'nil) (E) (A)))

(define (atom x) (if (pair? x) 'nil 't))
(define (car x) (if (pair? x) (s.car x) '()))
(define (cdr x) (if (pair? x) (s.cdr x) '()))
(define (equal x y) (if (equal? x y) 't 'nil))
(define (natp x)
  (if (integer? x) (if (< x 0) 'nil 't) 'nil))
(define (+ x y) (s.+ (num x) (num y)))
(define (< x y)
  (if (s.< (num x) (num y)) 't 'nil))
```

```
(define-syntax if
  (syntax-rules ()
    ((_ Q A E)
     (if/nil Q (lambda () A) (lambda () E)))))

(define-syntax defun
  (syntax-rules ()
    ((_ name (arg ...) body)
     (define (name arg ...) body))))

(define-syntax dethm
  (syntax-rules ()
    ((_ name (arg ...) body)
     (define (name arg ...) body))))

(defun size (x)
  (if (atom x)
    '0
    (+ '1 (size (car x)) (size (cdr x)))))
```

J-Bob. J-Bob, like ACL2, disallows forward references; therefore we write J-Bob in a bottom-up style. We adopt two naming conventions for functions. Constructors of tagged lists end in `-c`. Except for `equal`, `atom`, `natp`, and `<`, functions that return either true or false end in `?`.

Every theorem and function presented in the chapters of the book has been verified using ACL2 in addition to the step-by-step proofs we present using J-Bob.

The first definitions in our implementation of J-Bob are simple functions for list manipulation.

```
(defun list0 () '())
(defun list0? (x) (equal x '()))

(defun list1 (x) (cons x (list0)))
(defun list1? (x)
  (if (atom x) 'nil (list0? (cdr x))))
(defun elem1 (xs) (car xs))

(defun list2 (x y) (cons x (list1 y)))
(defun list2? (x)
  (if (atom x) 'nil (list1? (cdr x))))
(defun elem2 (xs) (elem1 (cdr xs)))

(defun list3 (x y z) (cons x (list2 y z)))
(defun list3? (x)
  (if (atom x) 'nil (list2? (cdr x))))
(defun elem3 (xs) (elem2 (cdr xs)))

(defun tag (sym x) (cons sym x))
(defun tag? (sym x)
  (if (atom x) 'nil (equal (car x) sym)))
(defun untag (x) (cdr x))

(defun member? (x ys)
  (if (atom ys)
    'nil
    (if (equal x (car ys))
      't
      (member? x (cdr ys)))))
```

We have chosen our data representations to mirror the syntax of the language of J-Bob. Our four kinds of expressions are `quote`, `if`, function application, and variable reference. This allows J-Bob to accept quoted expressions using the same syntax as J-Bob's definition. For defining functions and theorems, we use `defun` and `dethm`, respectively.

```
(defun quote-c (value)
  (tag 'quote (list1 value)))
(defun quote? (x)
  (if (tag? 'quote x) (list1? (untag x)) 'nil))
(defun quote.value (e) (elem1 (untag e)))

(defun if-c (Q A E) (tag 'if (list3 Q A E)))
(defun if? (x)
  (if (tag? 'if x) (list3? (untag x)) 'nil))
(defun if.Q (e) (elem1 (untag e)))
(defun if.A (e) (elem2 (untag e)))
(defun if.E (e) (elem3 (untag e)))

(defun app-c (name args) (cons name args))
(defun app? (x)
  (if (atom x)
    'nil
    (if (quote? x)
      'nil
      (if (if? x)
        'nil
        't))))
(defun app.name (e) (car e))
(defun app.args (e) (cdr e))

(defun var? (x)
  (if (equal x 't)
    'nil
    (if (equal x 'nil)
      'nil
      (if (natp x)
        'nil
        (atom x)))))

(defun defun-c (name formals body)
  (tag 'defun (list3 name formals body)))
(defun defun? (x)
  (if (tag? 'defun x) (list3? (untag x)) 'nil))
(defun defun.name (def) (elem1 (untag def)))
(defun defun.formals (def) (elem2 (untag def)))
(defun defun.body (def) (elem3 (untag def)))

(defun dethm-c (name formals body)
  (tag 'dethm (list3 name formals body)))
(defun dethm? (x)
  (if (tag? 'dethm x) (list3? (untag x)) 'nil))
(defun dethm.name (def) (elem1 (untag def)))
(defun dethm.formals (def) (elem2 (untag def)))
(defun dethm.body (def) (elem3 (untag def)))
```

The `if-QAE` and `QAE-if` functions convert between an `if` expression and a list of its three subexpressions. The `rator?` function recognizes operators and `rator.formals` produces the formal argument list of each.

```
(defun if-QAE (e)
  (list3 (if.Q e) (if.A e) (if.E e)))
(defun QAE-if (es)
  (if-c (elem1 es) (elem2 es) (elem3 es)))

(defun rator? (name)
  (member? name
    '(equal atom car cdr cons natp size + <)))

(defun rator.formals (rator)
  (if (member? rator '(atom car cdr natp size))
    '(x)
    (if (member? rator '(equal cons + <))
      '(x y)
      'nil)))
```

The `def.name` and `def.formals` functions extract parts of values that may be either a `defun` or a `dethm`.

```
(defun def.name (def)
  (if (defun? def)
    (defun.name def)
    (if (dethm? def)
      (dethm.name def)
      def)))

(defun def.formals (def)
  (if (dethm? def)
    (dethm.formals def)
    (if (defun? def)
      (defun.formals def)
      '())))
```

The function `if-c-when-necessary` constructs an `if` when its answer and else are distinct; `conjunction` and `implication` combine lists of expressions using appropriate nested `if`s.

```
(defun if-c-when-necessary (Q A E)
  (if (equal A E) A (if-c Q A E)))

(defun conjunction (es)
  (if (atom es)
    (quote-c 't)
    (if (atom (cdr es))
      (car es)
      (if-c (car es)
        (conjunction (cdr es))
        (quote-c 'nil)))))

(defun implication (es e)
  (if (atom es)
    e
    (if-c (car es)
      (implication (cdr es) e)
      (quote-c 't))))
```

The next group of functions manipulate sets and association lists. These association lists are represented as lists of definitions; `lookup` finds a definition in a list by comparing a given name to the `def.name` of each list element. The function `undefined?` reports when none of the definitions in a list match a given name.

The `args-arity?` and `app-arity?` functions determine whether an application has the right number of arguments.

```
(defun args-arity? (def args)
  (if (dethm? def)
    'nil
    (if (defun? def)
      (arity? (defun.formals def) args)
      (if (rator? def)
        (arity? (rator.formals def) args)
        'nil))))

(defun app-arity? (defs app)
  (args-arity? (lookup (app.name app) defs)
    (app.args app)))
```

```
(defun lookup (name defs)
  (if (atom defs)
    name
    (if (equal (def.name (car defs)) name)
      (car defs)
      (lookup name (cdr defs)))))

(defun undefined? (name defs)
  (if (var? name)
    (equal (lookup name defs) name)
    'nil))
```

J-Bob makes sure its inputs are sensible before proceeding with a proof; the functions `expr?` and `exprs?` check expressions and lists of expressions, respectively, to make sure that undefined names are not used, applications have the right number of arguments, etc. If `'any` is passed for the list of bound variables, J-Bob allows unbound variables; function and theorem bodies must only use bound variables, while proof steps may freely introduce new variables.

```
(defun bound? (var vars)
  (if (equal vars 'any) 't (member? var vars)))
```

```
(defun exprs? (defs vars es)
  (if (atom es)
    't
    (if (var? (car es))
      (if (bound? (car es) vars)
        (exprs? defs vars (cdr es))
        'nil)
      (if (quote? (car es))
        (exprs? defs vars (cdr es))
        (if (if? (car es))
          (if (exprs? defs vars
                (if-QAE (car es)))
            (exprs? defs vars (cdr es))
            'nil)
          (if (app? (car es))
            (if (app-arity? defs (car es))
              (if (exprs? defs vars
                    (app.args (car es)))
                (exprs? defs vars (cdr es))
                'nil)
              'nil)
            'nil))))))
(defun expr? (defs vars e)
  (exprs? defs vars (list1 e)))
```

Sets are represented as lists with no duplicates. Functions included are `subset?`, `list-extend`, and `list-union`, which also preserves the left-to-right ordering of elements in a predictable way that is convenient for constructing totality claims and inductive claims.

```
(defun subset? (xs ys)
  (if (atom xs)
    't
    (if (member? (car xs) ys)
      (subset? (cdr xs) ys)
      'nil)))

(defun list-extend (xs x)
  (if (atom xs)
    (list1 x)
    (if (equal (car xs) x)
      xs
      (cons (car xs)
        (list-extend (cdr xs) x)))))

(defun list-union (xs ys)
  (if (atom ys)
    xs
    (list-union (list-extend xs (car ys))
      (cdr ys))))
```

The functions `get-arg` and `set-arg` process argument lists indexed from 1.

```
(defun get-arg-from (n args from)
  (if (atom args)
    'nil
    (if (equal n from)
      (car args)
      (get-arg-from n (cdr args) (+ from '1)))))
(defun get-arg (n args)
  (get-arg-from n args '1))

(defun set-arg-from (n args y from)
  (if (atom args)
    '()
    (if (equal n from)
      (cons y (cdr args))
      (cons (car args)
        (set-arg-from n (cdr args) y
          (+ from '1))))))
(defun set-arg (n args y)
  (set-arg-from n args y '1))
```

Several predicates operate on argument lists: `<=len` reports whether a given index is less than or equal to the length of an argument list; `arity?` reports whether a list of formal arguments and a list of actual arguments have the same length; and `formals?` reports whether a given list represents distinct formal arguments.

```
(defun <=len-from (n args from)
  (if (atom args)
    'nil
    (if (equal n from)
      't
      (<=len-from n (cdr args) (+ from '1)))))
(defun <=len (n args)
  (if (< '0 n) (<=len-from n args '1) 'nil))

(defun arity? (vars es)
  (if (atom vars)
    (atom es)
    (if (atom es)
      'nil
      (arity? (cdr vars) (cdr es)))))

(defun formals? (vars)
  (if (atom vars)
    't
    (if (var? (car vars))
      (if (member? (car vars) (cdr vars))
        'nil
        (formals? (cdr vars)))
      'nil)))
```

The path to a focus is recognized by `path?` and `direction?`.

```
(defun direction? (dir)
  (if (natp dir)
    't
    (member? dir '(Q A E))))

(defun path? (path)
  (if (atom path)
    't
    (if (direction? (car path))
      (path? (cdr path))
      'nil)))
```

The function `quoted-exprs?` is a predicate that recognizes lists of quoted literals.

```
(defun quoted-exprs? (args)
  (if (atom args)
    't
    (if (quote? (car args))
      (quoted-exprs? (cdr args))
      'nil)))
```

The arguments to a proof step's application are checked by `step-args?`, which allows free variables but requires the arguments to an operator to be quoted literals.

```
(defun step-args? (defs def args)
  (if (dethm? def)
    (if (arity? (dethm.formals def) args)
      (exprs? defs 'any args)
      'nil)
    (if (defun? def)
      (if (arity? (defun.formals def) args)
        (exprs? defs 'any args)
        'nil)
      (if (rator? def)
        (if (arity? (rator.formals def) args)
          (quoted-exprs? args)
          'nil)
        'nil))))

(defun step-app? (defs app)
  (step-args? defs
    (lookup (app.name app) defs)
    (app.args app)))
```

The `steps?` function checks the syntax of proof steps to make sure they each contain a path and an application of a theorem or function name to the right number of syntactically valid expressions. In the case of operators, the arguments must be `quote` expressions.

```
(defun step? (defs step)
  (if (path? (elem1 step))
    (if (app? (elem2 step))
      (step-app? defs (elem2 step))
      'nil)
    'nil))

(defun steps? (defs steps)
  (if (atom steps)
    't
    (if (step? defs (car steps))
      (steps? defs (cdr steps))
      'nil)))
```

The *seed* of a proof is the induction scheme for a theorem's proof or a measure for a function's totality proof. An induction scheme must be an application of a defined function to a sequence of distinct variables, as checked by `induction-scheme?`. A measure may be any syntactically valid expression that refers only to previously defined functions and the formal arguments of the function being defined, which are checked by `seed?`.

```
(defun induction-scheme-for? (def vars e)
  (if (defun? def)
    (if (arity? (defun.formals def) (app.args e))
      (if (formals? (app.args e))
        (subset? (app.args e) vars)
        'nil)
      'nil)
    'nil))

(defun induction-scheme? (defs vars e)
  (if (app? e)
    (induction-scheme-for?
      (lookup (app.name e) defs)
      vars
      e)
    'nil))

(defun seed? (defs def seed)
  (if (equal seed 'nil)
    't
    (if (defun? def)
      (expr? defs (defun.formals def) seed)
      (if (dethm? def)
        (induction-scheme? defs
          (dethm.formals def)
          seed)
        'nil))))
```

Individual definitions are checked by **def?**, which checks that **defuns** and **dethms** have unique names, valid formal argument lists, and syntactically valid body expressions.

```
(defun extend-rec (defs def)
  (if (defun? def)
    (list-extend defs
      (defun-c
        (defun.name def)
        (defun.formals def)
        (app-c (defun.name def)
          (defun.formals def))))
    defs))

(defun def-contents? (known-defs formals body)
  (if (formals? formals)
    (expr? known-defs formals body)
    'nil))

(defun def? (known-defs def)
  (if (dethm? def)
    (if (undefined? (dethm.name def)
          known-defs)
      (def-contents? known-defs
        (dethm.formals def)
        (dethm.body def))
      'nil)
    (if (defun? def)
      (if (undefined? (defun.name def)
            known-defs)
        (def-contents?
          (extend-rec known-defs def)
          (defun.formals def)
          (defun.body def))
        'nil)
      'nil)))
```

```
(defun defs? (known-defs defs)
  (if (atom defs)
    't
    (if (def? known-defs (car defs))
      (defs? (list-extend known-defs (car defs))
        (cdr defs))
      'nil)))
```

J-Bob checks lists of proofs using **proofs?**, which in turn checks the definitions, seeds, and proof steps of each one using the previously described functions.

```
(defun list2-or-more? (pf)
  (if (atom pf)
    'nil
    (if (atom (cdr pf))
      'nil
      't)))

(defun proof? (defs pf)
  (if (list2-or-more? pf)
    (if (def? defs (elem1 pf))
      (if (seed? defs (elem1 pf) (elem2 pf))
        (steps? (extend-rec defs (elem1 pf))
          (cdr (cdr pf)))
        'nil)
      'nil)
    'nil))
```

We define substitution in **sub-e** and its helpers, which maintain separate, corresponding lists of formal arguments and the actual arguments to replace them with.

```
(defun proofs? (defs pfs)
  (if (atom pfs)
    't
    (if (proof? defs (car pfs))
      (proofs?
        (list-extend defs (elem1 (car pfs)))
        (cdr pfs))
      'nil)))

(defun sub-var (vars args var)
  (if (atom vars)
    var
    (if (equal (car vars) var)
      (car args)
      (sub-var (cdr vars) (cdr args) var))))
```

```
(defun sub-es (vars args es)
  (if (atom es)
    '()
    (if (var? (car es))
      (cons (sub-var vars args (car es))
        (sub-es vars args (cdr es)))
      (if (quote? (car es))
        (cons (car es)
          (sub-es vars args (cdr es)))
        (if (if? (car es))
          (cons
            (QAE-if
              (sub-es vars args
                (if-QAE (car es))))
            (sub-es vars args (cdr es)))
          (cons
            (app-c (app.name (car es))
              (sub-es vars args
                (app.args (car es))))
            (sub-es vars args (cdr es))))))))
(defun sub-e (vars args e)
  (elem1 (sub-es vars args (list1 e))))
```

The expr-recs and exprs-recs functions extract the recursive applications from the body of a function definition, given the function's name and the expression(s) in question.

```
(defun exprs-recs (f es)
  (if (atom es)
    '()
    (if (var? (car es))
      (exprs-recs f (cdr es))
      (if (quote? (car es))
        (exprs-recs f (cdr es))
        (if (if? (car es))
          (list-union
            (exprs-recs f (if-QAE (car es)))
            (exprs-recs f (cdr es)))
          (if (equal (app.name (car es)) f)
            (list-union
              (list1 (car es))
              (list-union
                (exprs-recs f
                  (app.args (car es)))
                (exprs-recs f (cdr es))))
            (list-union
              (exprs-recs f (app.args (car es)))
              (exprs-recs f
                (cdr es)))))))))
(defun expr-recs (f e)
  (exprs-recs f (list1 e)))
```

The function totality/claim and its corresponding function induction/claim are responsible for constructing totality claims and inductive claims, respectively. These functions construct their claims following the steps described in chapters 8 and 9, extended to account for recursive applications in the question of an if. These definitions rely on the predictable ordering of elements produced by list-union.

```
(defun totality/< (meas formals app)
  (app-c '<
    (list2 (sub-e formals (app.args app) meas)
      meas)))

(defun totality/meas (meas formals apps)
  (if (atom apps)
    '()
    (cons
      (totality/< meas formals (car apps))
      (totality/meas meas formals (cdr apps)))))

(defun totality/if (meas f formals e)
  (if (if? e)
    (conjunction
      (list-extend
        (totality/meas meas formals
          (expr-recs f (if.Q e)))
        (if-c-when-necessary (if.Q e)
          (totality/if meas f formals
            (if.A e))
          (totality/if meas f formals
            (if.E e)))))
    (conjunction
      (totality/meas meas formals
        (expr-recs f e)))))

(defun totality/claim (meas def)
  (if (equal meas 'nil)
    (if (equal (expr-recs (defun.name def)
                 (defun.body def))
               '())
      (quote-c 't)
      (quote-c 'nil))
    (if-c
      (app-c 'natp (list1 meas))
      (totality/if meas (defun.name def)
        (defun.formals def)
        (defun.body def))
      (quote-c 'nil))))
```

```
(defun induction/prems (vars claim apps)
  (if (atom apps)
    '()
    (cons
      (sub-e vars (app.args (car apps)) claim)
      (induction/prems vars claim (cdr apps)))))

(defun induction/if (vars claim f e)
  (if (if? e)
    (implication
      (induction/prems vars claim
        (expr-recs f (if.Q e)))
      (if-c-when-necessary (if.Q e)
        (induction/if vars claim f (if.A e))
        (induction/if vars claim f (if.E e))))
    (implication
      (induction/prems vars claim
        (expr-recs f e))
      claim)))

(defun induction/defun (vars claim def)
  (induction/if vars claim (defun.name def)
    (sub-e (defun.formals def) vars
      (defun.body def))))

(defun induction/claim (defs seed def)
  (if (equal seed 'nil)
    (dethm.body def)
    (induction/defun (app.args seed)
      (dethm.body def)
      (lookup (app.name seed) defs))))
```

Proof steps find the focus of a claim and the premises based on a *path* constructed as a list of *directions*: `'Q`, `'A`, `'E`, or a positive natural number, just like a walking path is constructed from the directions North, South, East, and West. Now that we have reached the *final* focus, we consider the second argument of the proof step: `(name arg ...)`. Next we substitute the variables of the associated conclusion—either a theorem body or a rewrite associated with applying a function—with the arguments `(arg ...)`, thus producing an *instantiated* conclusion.

Before we are permitted to use this instantiated theorem, we must check the premises with `follow-prems` against the instantiated theorem. This leads to an equality as the instantiated conclusion that represents two provably equal expressions.

```
(defun find-focus-at-direction (dir e)
  (if (equal dir 'Q)
    (if.Q e)
    (if (equal dir 'A)
      (if.A e)
      (if (equal dir 'E)
        (if.E e)
        (get-arg dir (app.args e))))))

(defun rewrite-focus-at-direction (dir e1 e2)
  (if (equal dir 'Q)
    (if-c e2 (if.A e1) (if.E e1))
    (if (equal dir 'A)
      (if-c (if.Q e1) e2 (if.E e1))
      (if (equal dir 'E)
        (if-c (if.Q e1) (if.A e1) e2)
        (app-c (app.name e1)
          (set-arg dir (app.args e1) e2))))))

(defun focus-is-at-direction? (dir e)
  (if (equal dir 'Q)
    (if? e)
    (if (equal dir 'A)
      (if? e)
      (if (equal dir 'E)
        (if? e)
        (if (app? e)
          (<=len dir (app.args e))
          'nil)))))

(defun focus-is-at-path? (path e)
  (if (atom path)
    't
    (if (focus-is-at-direction? (car path) e)
      (focus-is-at-path? (cdr path)
        (find-focus-at-direction (car path) e))
      'nil)))

(defun find-focus-at-path (path e)
  (if (atom path)
    e
    (find-focus-at-path (cdr path)
      (find-focus-at-direction (car path) e))))

(defun rewrite-focus-at-path (path e1 e2)
  (if (atom path)
    e2
    (rewrite-focus-at-direction (car path) e1
      (rewrite-focus-at-path (cdr path)
        (find-focus-at-direction (car path) e1)
        e2))))
```

```
(defun prem-A? (prem path e)
  (if (atom path)
    'nil
    (if (equal (car path) 'A)
      (if (equal (if.Q e) prem)
        't
        (prem-A? prem (cdr path)
          (find-focus-at-direction (car path)
            e)))
      (prem-A? prem (cdr path)
        (find-focus-at-direction (car path)
          e)))))

(defun prem-E? (prem path e)
  (if (atom path)
    'nil
    (if (equal (car path) 'E)
      (if (equal (if.Q e) prem)
        't
        (prem-E? prem (cdr path)
          (find-focus-at-direction (car path)
            e)))
      (prem-E? prem (cdr path)
        (find-focus-at-direction (car path)
          e)))))

(defun follow-prems (path e thm)
  (if (if? thm)
    (if (prem-A? (if.Q thm) path e)
      (follow-prems path e (if.A thm))
      (if (prem-E? (if.Q thm) path e)
        (follow-prems path e (if.E thm))
        thm))
    thm))
```

The function `apply-op` applies an operator to a list of values and returns a value.

```
(defun unary-op (rator rand)
  (if (equal rator 'atom)
    (atom rand)
    (if (equal rator 'car)
      (car rand)
      (if (equal rator 'cdr)
        (cdr rand)
        (if (equal rator 'natp)
          (natp rand)
          (if (equal rator 'size)
            (size rand)
            'nil)))))))
```

```
(defun binary-op (rator rand1 rand2)
  (if (equal rator 'equal)
    (equal rand1 rand2)
    (if (equal rator 'cons)
      (cons rand1 rand2)
      (if (equal rator '+)
        (+ rand1 rand2)
        (if (equal rator '<)
          (< rand1 rand2)
          'nil)))))

(defun apply-op (rator rands)
  (if (member? rator '(atom car cdr natp size))
    (unary-op rator (elem1 rands))
    (if (member? rator '(equal cons + <))
      (binary-op rator
        (elem1 rands)
        (elem2 rands))
      'nil)))
```

When a proof step's application refers to an operator, `eval-op` produces an expression representing the value of that application, and `rands` function extracts the values of the operator's quoted arguments.

```
(defun rands (args)
  (if (atom args)
    '()
    (cons (quote.value (car args))
      (rands (cdr args)))))

(defun eval-op (app)
  (quote-c
    (apply-op (app.name app)
      (rands (app.args app)))))
```

The proof steps rewrite the focus within a claim to one that is provably equal to it. There are three kinds of proof steps as is evident in the definition of `equality/def`. A proof step may use the operators (`equal`, `atom`, `car`, `cdr`, `cons`, `natp`, `size`, `+`, or `<`) applied to quoted values, yielding a quoted value; the definition of a function; or a possibly conditional equality in any defined theorem. In each proof step, J-Bob produces two expressions that may be rewritten to each other.

These two expressions are then passed along to `equality` with the final focus. This function checks whether either expression is equal to the final focus.

If so, `equality` performs the rewrite by returning the other expression as the new focus. If not, the proof step has failed and `equality` returns the final focus unchanged. In either case, we are left to expand out the context around this possibly new focus. Thus, as the recursion of `equality/path` cautiously unwinds into `rewrite-focus-at-direction`, we go backwards one direction at a time, rebuilding the original context yielding either a new claim with the new focus or the original claim.

```
(defun app-of-equal? (e)
  (if (app? e)
    (equal (app.name e) 'equal)
    'nil))

(defun equality (focus a b)
  (if (equal focus a)
    b
    (if (equal focus b)
      a
      focus)))

(defun equality/equation (focus concl-inst)
  (if (app-of-equal? concl-inst)
    (equality focus
      (elem1 (app.args concl-inst))
      (elem2 (app.args concl-inst)))
    focus))

(defun equality/path (e path thm)
  (if (focus-is-at-path? path e)
    (rewrite-focus-at-path path e
      (equality/equation
        (find-focus-at-path path e)
        (follow-prems path e thm)))
    e))
```

```
(defun equality/def (claim path app def)
  (if (rator? def)
    (equality/path claim path
      (app-c 'equal (list2 app (eval-op app))))
    (if (defun? def)
      (equality/path claim path
        (sub-e (defun.formals def)
          (app.args app)
          (app-c 'equal
            (list2
              (app-c (defun.name def)
                (defun.formals def))
              (defun.body def)))))
      (if (dethm? def)
        (equality/path claim path
          (sub-e (dethm.formals def)
            (app.args app)
            (dethm.body def)))
        claim))))
```

The function `rewrite/steps` rewrites the given claim according to each proof step in the list of suggested proof steps, stopping either when there are no more proof steps or when a proof step fails. A proof step is considered to fail when it rewrites a claim to the same claim. The function `rewrite/continue` passes the *new* claim that results from the first proof step to `rewrite/steps` and the *old* claim.

```
(defun rewrite/step (defs claim step)
  (equality/def claim (elem1 step) (elem2 step)
    (lookup (app.name (elem2 step)) defs)))

(defun rewrite/continue (defs steps old new)
  (if (equal new old)
    new
    (if (atom steps)
      new
      (rewrite/continue defs (cdr steps) new
        (rewrite/step defs new (car steps))))))

(defun rewrite/steps (defs claim steps)
  (if (atom steps)
    claim
    (rewrite/continue defs (cdr steps) claim
      (rewrite/step defs claim (car steps)))))
```

```
(defun rewrite/prove (defs def seed steps)
  (if (defun? def)
    (rewrite/steps defs
      (totality/claim seed def)
      steps)
    (if (dethm? def)
      (rewrite/steps defs
        (induction/claim defs seed def)
        steps)
      (quote-c 'nil))))

(defun rewrite/prove+1 (defs pf e)
  (if (equal e (quote-c 't))
    (rewrite/prove defs (elem1 pf) (elem2 pf)
      (cdr (cdr pf)))
    e))

(defun rewrite/prove+ (defs pfs)
  (if (atom pfs)
    (quote-c 't)
    (rewrite/prove+1 defs (car pfs)
      (rewrite/prove+
        (list-extend defs (elem1 (car pfs)))
        (cdr pfs)))))
```

```
(defun rewrite/define (defs def seed steps)
  (if (equal (rewrite/prove defs def seed steps)
             (quote-c 't))
    (list-extend defs def)
    defs))

(defun rewrite/define+1 (defs1 defs2 pfs)
  (if (equal defs1 defs2)
    defs1
    (if (atom pfs)
      defs2
      (rewrite/define+1 defs2
        (rewrite/define defs2
          (elem1 (car pfs))
          (elem2 (car pfs))
          (cdr (cdr (car pfs))))
        (cdr pfs)))))

(defun rewrite/define+ (defs pfs)
  (if (atom pfs)
    defs
    (rewrite/define+1 defs
      (rewrite/define defs
        (elem1 (car pfs))
        (elem2 (car pfs))
        (cdr (cdr (car pfs))))
      (cdr pfs))))
```

Our proof assistant, J-Bob, has an interactive front-end: `J-Bob/prove`. It accepts a list of established definitions (theorems or functions) and a list of *proof attempts*. Each proof attempt is a list of two or more elements: a proposed definition (`defun` or `dethm`), a *seed* (measure expression or induction scheme), and zero or more *proof steps*. A proof step specifies a justification for an individual rewrite contributing to a proof that the proposed theorem is true or that the proposed function is total.

`J-Bob/prove` processes proof attempts in reverse order, attempting the last one first, similarly to how we prove **align**'s totality claim first in chapter 10 before returning to **natp/wt** and **positive/wt**. If a proof fails at any point, `rewrite/prove+` stops and produces the result of the last correct rewrite. Otherwise it produces the final quoted expression `''t`.

`J-Bob/define` is designed to extend an existing set of definitions with a list of definitions if the corresponding proof attempts succeed. Finally, `J-Bob/step` processes a sequence of rewriting steps in order without any proof attempt. As in the recess of chapter 2, this can be used to explore rewriting without an explicit goal of `''t`.

```
(defun J-Bob/step (defs e steps)
  (if (defs? '() defs)
    (if (expr? defs 'any e)
      (if (steps? defs steps)
        (rewrite/steps defs e steps)
        e)
      e)
    e))

(defun J-Bob/prove (defs pfs)
  (if (defs? '() defs)
    (if (proofs? defs pfs)
      (rewrite/prove+ defs pfs)
      (quote-c 'nil))
    (quote-c 'nil)))

(defun J-Bob/define (defs pfs)
  (if (defs? '() defs)
    (if (proofs? defs pfs)
      (rewrite/define+ defs pfs)
      defs)
    defs))
```

```
(defun axioms ()
  '((dethm atom/cons (x y)
      (equal (atom (cons x y)) 'nil))
    (dethm car/cons (x y)
      (equal (car (cons x y)) x))
    (dethm cdr/cons (x y)
      (equal (cdr (cons x y)) y))
    (dethm equal-same (x)
      (equal (equal x x) 't))
    (dethm equal-swap (x y)
      (equal (equal x y) (equal y x)))
    (dethm if-same (x y)
      (equal (if x y y) y))
    (dethm if-true (x y)
      (equal (if 't x y) x))
    (dethm if-false (x y)
      (equal (if 'nil x y) y))
    (dethm if-nest-E (x y z)
      (if x 't (equal (if x y z) z)))
    (dethm if-nest-A (x y z)
      (if x (equal (if x y z) y) 't))
    (dethm cons/car+cdr (x)
      (if (atom x)
        't
        (equal (cons (car x) (cdr x)) x)))
    (dethm equal-if (x y)
      (if (equal x y) (equal x y) 't))
    (dethm natp/size (x)
      (equal (natp (size x)) 't))
    (dethm size/car (x)
      (if (atom x)
        't
        (equal (< (size (car x)) (size x)) 't)))
    (dethm size/cdr (x)
      (if (atom x)
        't
        (equal (< (size (cdr x)) (size x)) 't)))
    (dethm associate-+ (a b c)
      (equal (+ (+ a b) c) (+ a (+ b c))))
    (dethm commute-+ (x y)
      (equal (+ x y) (+ y x)))
    (dethm natp/+ (x y)
      (if (natp x)
        (if (natp y)
          (equal (natp (+ x y)) 't)
          't)
        't))
    (dethm positives-+ (x y)
      (if (< '0 x)
        (if (< '0 y)
          (equal (< '0 (+ x y)) 't)
          't)
        't))
    (dethm common-addends-< (x y z)
      (equal (< (+ x z) (+ y z)) (< x y)))
    (dethm identity-+ (x)
      (if (natp x) (equal (+ '0 x) x) 't))))
```

The function **axioms** produces a list of the axioms we use in the preceding chapters, and **prelude** extends it with the definitions and proofs of totality claims of **list-induction** and **star-induction** used for List Induction and Star Induction, respectively.

```
(defun prelude ()
  (J-Bob/define (axioms)
    '(((defun list-induction (x)
         (if (atom x)
           '()
           (cons (car x)
             (list-induction (cdr x)))))
       (size x)
       ((A E) (size/cdr x))
       ((A) (if-same (atom x) 't))
       ((Q) (natp/size x))
       (() (if-true 't 'nil)))
      ((defun star-induction (x)
         (if (atom x)
           x
           (cons (star-induction (car x))
             (star-induction (cdr x)))))
       (size x)
       ((A E A) (size/cdr x))
       ((A E Q) (size/car x))
       ((A E) (if-true 't 'nil))
       ((A) (if-same (atom x) 't))
       ((Q) (natp/size x))
       (() (if-true 't 'nil))))))
```

Restless for More?

We hope that you have enjoyed reading this book, and have enjoyed playing with our proof assistant J-Bob. If you want to find out more about proofs and logic, there are several options for where to turn next.

J-Bob is named after J Moore and Bob Boyer, who wrote one of the earliest mechanical theorem provers called Nqthm. The modern version of Nqthm, written by J Moore and Matt Kaufmann, is called ACL2. ACL2 allows you to play with theorems like you can in J-Bob, but does a lot of the "heavy lifting" for you. That way you can prove more complicated theorems with less effort. In fact, ACL2 can prove every totality claim and theorem in this book given only the relevant `defun`s and `dethm`s. It does all the rest of the work for you.

ACL2 can be found at `http://www.cs.utexas.edu/users/moore/acl2/`. For a simple user interface to ACL2, try "Dracula," available at `http://dracula-lang.org/`, which adds tools such as syntax checking, automated testing, and the ability to prove theorems about programs that use pictures and animations. For access to more powerful ACL2 features, check out the "ACL2 Sedan" at `http://acl2s.ccs.neu.edu/acl2s/doc/`, which adds even more automation to totality proofs and supports the full range of ACL2's features.

J-Bob and ACL2 are not the only kind of mechanical theorem provers that exist. There are many others that support different languages of expressions. Examples include Agda (`http://wiki.portal.chalmers.se/agda/`), Coq (`http://coq.inria.fr/`), Isabelle/HOL (`http://www.cl.cam.ac.uk/research/hvg/Isabelle/`), PVS (`http://pvs.csl.sri.com/`), and Twelf (`http://twelf.org/`).

We can also recommend some further reading on the topic of proofs and logic. The modern founder of the use of recursion in Logic is Thoralf Albert Skolem. In Skolem's 1919 paper (published in 1923), he observed that one could use *the recursive mode of thought* to avoid "some" used in Whitehead and Russell's *Principia Mathematica*. The ideas of this paper were important to the development of the Boyer-Moore Theorem Prover. Among other books that we think you might find worthwhile are the books on Logic and Mathematics mentioned below. Not all the books are easy to find or to read, but most have survived the test of time.

R. S. Boyer and J S. Moore. *A Computational Logic.* Academic Press, Inc., New York, 1979.

A. Chlipala. *Certified Programming with Dependent Types.* MIT Press, 2013.

J. N. Crossley, C. J. Ash, C. J. Brickhill, J. C. Stillwell, and N. H. Williams. *What is Mathematical Logic?* Oxford University Press, 1972.

M. Kaufmann, P. Manolios, and J S. Moore. *Computer Aided Reasoning: An Approach.* Kluwer Academic Publishers, 2000.

D. MacKenzie. *Mechanizing Proof: Computing, Risk, and Trust.* MIT Press, 2004.

J. McCarthy. A Basis for a Mathematical Theory of Computation. In P. Braffort and D. Hershberg (Eds.), *Computer Programming and Formal Systems.* North-Holland Publishing Company, Amsterdam, The Netherlands, 1963.

E. Mendelson. *Introduction to Mathematical Logic.* D. Van Nostrand Company, Inc., Princeton, New Jersey, 1964.

R. Péter. *Recursive Functions Third Revised Edition.* Academic Press, New York, 1967.

Pierce, B. C., et al. *Software Foundations.* `http://www.cis.upenn.edu/~bcpierce/sf` (2010-2015).

T. A. Skolem. The foundations of elementary arithmetic established by means of the recursive mode of thought, without the use of apparent variables ranging over infinite domains, in *From Frege to Gödel: A Source Book in Mathematical Logic, 1879–1931* (Jean van Heijenoort, ed.), pages 302–333. Harvard Univ. Press, 1967. Paper written in 1919 and appeared in published form in 1923.

P. Suppes. *Introduction to Logic.* D. Van Nostrand Company, Inc., Princeton, New Jersey, 1957.

M. Wand. *Induction, Recursion, and Programming.* Elsevier North Holland, Inc., 1980.

A. N. Whitehead and B. Russell. *Principia Mathematica.* Cambridge: Cambridge University Press, in 3 vols, 1910, 1912, 1913. Second edition, 1925 (Vol. 1), 1927 (Vols 2, 3). Abridged as *Principia Mathematica to *56*, Cambridge University Press, 1962.

Afterword

In January 1984, I went to Indiana to get my PhD from Dan Friedman. Among other ideas, he was well known for *The Little LISPer*, which everybody used to teach recursion. Back then, recursion seemed so complicated, but Dan's book helped everyone understand it.

A year later, Dan put me to work on a revision of *The Little LISPer*. It had aged a bit since its first appearance in 1974, and I had proposed some small improvements that he liked. We spent nine long months working it all out. For me, as a second year graduate student, it was an exhilarating experience. Working with Dan, I learned a lot about how to take a research insight and think it over until the explanation was accessible to freshman students.

Fast forward. In the spring of 2008, I proposed a novel kind of freshman course that would explain the role of logic and theorem proving in programming. I asked two of my PhD students, Dale Vaillancourt and Carl Eastlund, to join me. Our goal was to develop a course in which students would first program functions like `sub`, `add-atoms`, plus some small video games and then make up and prove theorems about them. We used the ACL2 proof assistant, a brilliant system for true experts, but it became clear quickly that freshman students needed a simple way to interact with this complex system.

Carl took on the task of creating an interactive development environment in which students could create programs and proofs together. We also continued to think about how to teach the material. I eventually realized that we needed Dan's way of studying, twisting, refining this material. So I discussed the project with him, and he suggested the *Little* book approach. But *Little* books have always been Dan's forte, and Carl had acquired a deeper understanding of ACL2 than most people on the planet. What would be more natural than for the two of them, for "grandfather" and "grandson," to co-author a *Little* book?

The Little Prover is the result of this collaboration. Like all of its predecessors, the book distills a topic to its essential elements. Here you find out about the mechanical method of using structural induction in proofs, which is critical for all current applications of logic to programming, and the J-Bob proof assistant, which can check such proofs. Study this book, work through it, and you will comprehend. And all you need to get through the book, is an understanding of *The Little LISPer*.

Now I have come full circle. I enjoyed the journey, and I hope you did, too.

Matthias Felleisen
Boston, Massachusetts

Index

Index

', *see* quote
(), 4
., *see* dotted pair
<=len, *see* J-Bob
<=len-from, *see* J-Bob

A, *see* if answer
ACL2, 203
add-atoms, *see* function
align, *see* function
align/align, *see* theorem
answer, *see* if answer
app-arity?, *see* J-Bob
app-c, *see* J-Bob
app-of-equal?, *see* J-Bob
app.args, *see* J-Bob
app.name, *see* J-Bob
app?, *see* J-Bob
apply-op, *see* J-Bob
args-arity?, *see* J-Bob
arity?, *see* J-Bob
associate-+, *see* axiom
atom, *see* operator
atom/cons, *see* axiom
atoms, *see* function
attention, 20
axiom, 7, **7, 8**
- associate-+, **148**
- atom/cons, 5, **8**, **26**
- car/cons, 6, **8**, **12**, **26**
- cdr/cons, 6, **8**, **26**
- common-addends-<, **148**
- commute-+, **148**
- cons/car+cdr, 25, **26**
- equal-if, 17, **18**
- equal-same, 9, **10**, **18**
  - versus equal, 17
- equal-swap, 9, **10**, **18**
- identity-+, **148**
- if-false, **15**, **27**
- if-nest-A, **27**, 29
- if-nest-E, **27**, 30
- if-same, **15**, **27**
  - in reverse, 16
- if-true, **15**, 17, **27**
- natp/+, **148**
- natp/size, **53**
- positives-+, **148**
- size/car, **53**
- size/cdr, **53**

axioms, *see* J-Bob
Axioms of
- + and <, **148**
- Cons, **8**, **26**
- Equal, **10**, **18**
- If, **15**, **27**
- Size, **53**

binary-op, *see* J-Bob
black, 5
blue, 5
bound?, *see* J-Bob

car, *see* operator
car/cons, *see* axiom
cdr, *see* operator

cdr/cons, *see* axiom
claim, **33**, 34
    totality, *see* totality claim
common-addends-<, *see* axiom
commute-+, *see* axiom
conclusion, **18**, 19
condition, **18**, 22–24
conjunction, **111**, 113
conjunction, *see* J-Bob
cons, *see* operator
cons/car+cdr, *see* axiom
Constructing Totality Claims, **113**
context, 5
contradiction, *see* theorem
counterexample, 118, 144, 157
ctx?, *see* function
ctx?/sub, *see* theorem
ctx?/t, *see* theorem

def-contents?, *see* J-Bob
def.formals, *see* J-Bob
def.name, *see* J-Bob
def?, *see* J-Bob
defs?, *see* J-Bob
defun, xii, *see also* The Law of Defun
Defun Induction, **124**, 118–127, 157
defun-c, *see* J-Bob
defun.body, *see* J-Bob
defun.formals, *see* J-Bob
defun.name, *see* J-Bob
defun?, *see* J-Bob
dethm, xiii, **8**, 12, *see also* The Law of Dethm
dethm-c, *see* J-Bob
dethm.body, *see* J-Bob
dethm.formals, *see* J-Bob
dethm.name, *see* J-Bob
dethm?, *see* J-Bob
direction?, *see* J-Bob
dotted pair, **115**
dotted pairs, 139–141, 145–146

E, *see* if else
elem1, *see* J-Bob
elem2, *see* J-Bob
elem3, *see* J-Bob
else, *see* if else
equal, 3
    both directions, 10
equal, *see* operator
equal-if, *see* axiom
equal-same, *see* axiom
equal-swap, *see* axiom
equality, *see* J-Bob
equality/def, *see* J-Bob
equality/equation, *see* J-Bob
equality/path, *see* J-Bob
eval-op, *see* J-Bob
evaluate, xi
expr-recs, *see* J-Bob
expr?, *see* J-Bob
expression, xii, 3
exprs-recs, *see* J-Bob
exprs?, *see* J-Bob
extend-rec, *see* J-Bob

find-focus-at-direction, *see* J-Bob
find-focus-at-path, *see* J-Bob
first-of, *see* function
first-of-pair, *see* theorem
focus, 5
    whole expression, 6
focus-is-at-direction?, *see* J-Bob
focus-is-at-path?, *see* J-Bob
follow-prems, *see* J-Bob
food, xiii
formals?, *see* J-Bob
function, xii, *see also* operator
    add-atoms, **108**–113, 115, 192
    align, **142**, 141–145, 147–156, 197, 200
    application, xii
    atom, 45
    atoms, **108**, 115, 193
    built-in, xii
    ctx?, **89**–90, 188
    equal, 43
    first-of, **33**, 183

in-pair?, **36**
index2, 46
list-induction, **125**, 178
list0?, **43**–44, 184
list1?, **44**–45, 184
list2?, **46**–184
list3?, **50**
list?, **50**–54, 185
measure, 52
memb?, **59**, 60, 185
measure, 59
member?, **107**, 192
pair, **33**, 183
partial, **47**
totality claim, 57
remb, **59**, 60–61, 185
rembmeasure, 60
rotate, **140**, 139–140, 196
second-of, **33**, 183
set?, **107**, 108, 192
star-induction, **125**, 179
sub, **56**, 55–56, 90
sub?, 185
wt, **146**, 145–147, 197

get-arg, *see* J-Bob
get-arg-from, *see* J-Bob

identity-+, *see* axiom
if, xii, 15, 45, 94, *see also* Axioms of If
answer, xii, **16**
else, xii, **16**
question, xii, **16**
If Lifting, 28–31, 66–67, **67**, 72, 81, 92, 93, 98, 99, 103, 104, 126, 128, 130, 131
if-c, *see* J-Bob
if-c-when-necessary, *see* J-Bob
if-false, *see* axiom
if-nest-A, *see* axiom
if-nest-E, *see* axiom
if-QAE, *see* J-Bob
if-same, *see* axiom
if-true, *see* axiom

if.A, *see* J-Bob
if.E, *see* J-Bob
if.Q, *see* J-Bob
if?, *see* J-Bob
implication, **122**
implication, *see* J-Bob
in-first-of-pair, *see* theorem
in-pair?, *see* function
in-second-of-pair, *see* theorem
induction, xi, **77**
defun, *see* Defun Induction
on lists, *see* List Induction
on nested conses, *see* Star Induction
induction-scheme-for?, *see* J-Bob
induction-scheme?, *see* J-Bob
induction/claim, *see* J-Bob
induction/defun, *see* J-Bob
induction/if, *see* J-Bob
induction/prems, *see* J-Bob
inductive claim, 119–124
inductive premise, **79**, 82, 99, 100, 104, 117–118, 120–121, 124, 134, 157, 161, 163
constructing, **121**
Insight
Build Up to Induction Gradually, **86**
Combine Ifs, **92**
Don't Touch Inductive Premises, **83**
Helper Theorems for Induction, **95**
Helper Theorems for Repetition, **144**
Know Theorems and Axioms, **69**
Pull Ifs Outward, **68**
Rewrite from the Inside Out, **63**
Skip Irrelevant Expressions, **39**

J-Bob, xiii, 165–179, 203–215
<=len, **206**
<=len-from, **206**
app-arity?, **205**
app-c, **204**

app-of-equal?, **212**
app.args, **204**
app.name, **204**
app?, **204**
apply-op, **211**
args-arity?, **205**
arity?, **206**
axioms, **214**
binary-op, **211**
bound?, **205**
conjunction, **205**
def-contents?, **208**
def.formals, **205**
def.name, **205**
def?, **208**
defs?, **208**
defun-c, **204**
defun.body, **204**
defun.formals, **204**
defun.name, **204**
defun?, **204**
dethm-c, **204**
dethm.body, **204**
dethm.formals, **204**
dethm.name, **204**
dethm?, **204**
direction?, **207**
elem1, **204**
elem2, **204**
elem3, **204**
equality, **212**
equality/def, **212**
equality/equation, **212**
equality/path, **212**
eval-op, **211**
expr-recs, **209**
expr?, **206**
exprs-recs, **209**
exprs?, **206**
extend-rec, **208**
find-focus-at-direction, **210**
find-focus-at-path, **210**
focus-is-at-direction?, **210**
focus-is-at-path?, **210**
follow-prems, **211**
formals?, **206**
get-arg, **206**
get-arg-from, **206**
if-c, **204**
if-c-when-necessary, **205**
if-QAE, **205**
if.A, **204**
if.E, **204**
if.Q, **204**
if?, **204**
implication, **205**
induction-scheme-for?, **207**
induction-scheme?, **207**
induction/claim, **210**
induction/defun, **210**
induction/if, **210**
induction/prems, **210**
J-Bob/define, 173, **213**
J-Bob/prove, 171, **213**
J-Bob/step, 167, **213**
list-extend, **206**
list-union, **206**
list0, **204**
list0?, **204**
list1, **204**
list1?, **204**
list2, **204**
list2-or-more?, **208**
list2?, **204**
list3, **204**
list3?, **204**
lookup, **205**
member?, **204**
path?, **207**
prelude, **214**
prem-A?, **211**
prem-E?, **211**
proof?, **208**
proofs?, **208**
QAE-if, **205**
quote-c, **204**

quote.value, **204**
quote?, **204**
quoted-exprs?, **207**
rands, **211**
rator.formals, **205**
rator?, **205**
rewrite-focus-at-direction, **210**
rewrite-focus-at-path, **210**
rewrite/continue, **212**
rewrite/define, **213**
rewrite/define+, **213**
rewrite/define+1, **213**
rewrite/prove, **213**
rewrite/prove+, **213**
rewrite/prove+1, **213**
rewrite/step, **212**
rewrite/steps, **212**
seed?, **207**
set-arg, **206**
set-arg-from, **206**
step-app?, **207**
step-args?, **207**
step?, **207**
steps?, **207**
sub-e, **208**
sub-es, **208**
sub-var, **208**
subset?, **206**
tag, **204**
tag?, **204**
totality/<, **209**
totality/claim, **209**
totality/if, **209**
totality/meas, **209**
unary-op, **211**
undefined?, **205**
untag, **204**
var?, **204**
J-Bob/define, *see* J-Bob
J-Bob/prove, *see* J-Bob
J-Bob/step, *see* J-Bob
jabberwocky, **20**

The Law of Defun, **34**, **47**
The Law of Dethm, **12**, **18**
condition, *see* condition
list, xii, 43, **50**
circular, 50
improper, *see* dotted pair
infinite, 50
List Induction, 78, **86**, 124
list-extend, *see* J-Bob
list-induction, *see* function
list-union, *see* J-Bob
list0, *see* J-Bob
list0?, *see* function, *see also* J-Bob
list1, *see* J-Bob
list1?, *see* function, *see also* J-Bob
list2, *see* J-Bob
list2-or-more?, *see* J-Bob
list2?, *see* function, *see also* J-Bob
list3, *see* J-Bob
list3?, *see* function, *see also* J-Bob
list?, *see* function
literal, xii
The Little LISPer, 3
The Little Schemer, 3
lookup, *see* J-Bob

measure, 51, **52**, 109–113
memb?, *see* function
memb?/remb, *see* theorem
memb?/remb0, *see* theorem
memb?/remb1, *see* theorem
memb?/remb2, *see* theorem
memb?/remb3, *see* theorem
member?, *see* function, *see also* J-Bob

natp/+, *see* axiom
natp/size, *see* axiom
natp/wt, *see* theorem
natural number, xii, **53**, 113
natural recursion, **77**, 78, 79, 90, 117–118, 161
nil, 3
notation, xii

operator, 45

+, 146, *see also* Axioms of + and <
<, 113, *see also* Axioms of + and <
atom, 4, *see also* Axioms of Cons
car, 3, *see also* Axioms of Cons
cdr, 3, *see also* Axioms of Cons
cons, 3, *see also* Axioms of Cons
dotted, *see* dotted pair
equal, **8**, *see also* Axioms of Equal
versus equal-same, 17
natp, 53, 113
size, 52, *see also* Axioms of Size
orange, 20
ordinal number, 126

pair, *see* function
partial, *see* function
partial function, **47**
path?, *see* J-Bob
positive/wt, *see* theorem
positives-+, *see* axiom
prelude, 165
prelude, *see* J-Bob
prem-A?, *see* J-Bob
prem-E?, *see* J-Bob
premise, **20**
inductive, *see* inductive premise
proof, **34**
proof assistant, xiii
proof attempt, **171**
Proof by
Induction, 100
List Induction, *see* List Induction
Star Induction, *see* Star Induction
proof?, *see* J-Bob
proofs?, *see* J-Bob
prove, **34**

Q, *see* if question
QAE-if, *see* J-Bob
question, *see* if question
quote, xii, 3
quote-c, *see* J-Bob
quote.value, *see* J-Bob
quote?, *see* J-Bob
quoted-exprs?, *see* J-Bob

rands, *see* J-Bob
rator.formals, *see* J-Bob
rator?, *see* J-Bob
recursion, 50, 77, 79
natural, *see* natural recursion
recursive function, xi
remb, *see* function
reverse, xi
rewrite, 3
rewrite-focus-at-direction, *see* J-Bob
rewrite-focus-at-path, *see* J-Bob
rewrite/continue, *see* J-Bob
rewrite/define, *see* J-Bob
rewrite/define+, *see* J-Bob
rewrite/define+1, *see* J-Bob
rewrite/prove, *see* J-Bob
rewrite/prove+, *see* J-Bob
rewrite/prove+1, *see* J-Bob
rewrite/step, *see* J-Bob
rewrite/steps, *see* J-Bob
rotate, *see* function
rotate/cons, *see* theorem

Scheme, 203
second-of, *see* function
second-of-pair, *see* theorem
seed, **171**
induction
on lists, *see* list-induction
on nested conses, *see* star-induction
measure, 176
recursive function, 176
seed?, *see* J-Bob
set-arg, *see* J-Bob
set-arg-from, *see* J-Bob
set?, *see* function
set?/add-atoms, *see* theorem, *see* theorem
set?/atoms, *see* theorem

set?/nil, *see* theorem
set?/t, *see* theorem
size/car, *see* axiom
size/cdr, *see* axiom
Star Induction, **91**, 116, 124
star-induction, *see* function
step, 5, 34, 165
step-app?, *see* J-Bob
step-args?, *see* J-Bob
step?, *see* J-Bob
steps?, *see* J-Bob
sub, *see* function
sub-e, *see* J-Bob
sub-es, *see* J-Bob
sub-var, *see* J-Bob
subset?, *see* J-Bob
symbol, xii

t, 4
tag, *see* J-Bob
tag?, *see* J-Bob
theorem, xiii, **8**, 34
  align/align, **157**–163, 201
  contradiction, **47**–50, 184
  ctx?/sub, **90**–100, 190
  ctx?/t, **95**, 100–104, 190
  first-of-pair, **33**–35, 183
  helper, 95, 116, 119, 129, 131, 141, 147, 150
  if-false, 49
  in-first-of-pair, **36**–38, 184
  in-pair?, 183
  in-second-of-pair, **38**–40, 184
  memb?/remb, **78**–85, 187
  memb?/remb0, **61**–62, 186
  memb?/remb1, **63**–70, 186
  memb?/remb2, **71**–74, 187
  memb?/remb3, **75**
  natp/wt, **147**, 200
  positive/wt, **150**, 200
  rotate/cons, **140**–141, 197
  second-of-pair, **35**–36, 183
  set?/add-atoms, **116**–117, **119**–134, 136, 193, 195
  set?/atoms, **116**, 134–136, 193, 195
  set?/nil, **131**, 195
  set?/t, **129**, 195
total, **43**, 47
totality claim, 53, 108–113, *see also* Constructing Totality Claims
  non-recursive function, 55
totality/<, *see* J-Bob
totality/claim, *see* J-Bob
totality/if, *see* J-Bob
totality/meas, *see* J-Bob

unary-op, *see* J-Bob
undefined?, *see* J-Bob
untag, *see* J-Bob

value, 3
  finite, 50
var?, *see* J-Bob
variable, xii, 16

wt, *see* function